# Praise for *Th...*

D0462197

"History shows that the human bo ... ary range of different traditional diets. Alas, one of the very few diets to which we are *not* well adapted is the Western diet most of us are eating today. In this bracingly hopeful and eminently practical book, Daphne Miller shows us how we can bring the wisdom of traditional diets to our own plates, in the interest of both our health and our pleasure. *The Jungle Effect* is a fascinating, useful, and important book."

— Michael Pollan, #1 *New York Times* bestselling
author of *The Omnivore's Dilemma* and *In Defense of Food*

"Miller's work is consistently informative and educational ... and Miller's practical advice and recipes are all geared for the novice. Anyone unafraid of modifying their diet will find this anthropological diet guide useful."

— *Publishers Weekly* online

"*The Jungle Effect* was such an enjoyable read that I almost forgot I was being fed a steady dose of valuable nutrition advice—advice that combines the wisdom of our ancestors with the latest nutrition research."

— Bradley J. Willcox, MD, coauthor of the *New York Times* bestseller *The Okinawa Program* and clinician-scientist, Pacific Health Research Institute, University of Hawaii

"Daphne Miller is the Sherlock Holmes of healthy eating. *The Jungle Effect* is an odyssey where she follows clues and food experts to discover some of the healthiest diets around the world—and how best to recreate those meals and lifestyles in our daily lives."

—Juliette Rossant, author of *Super Chef*

"*The Jungle Effect* is a brilliant piece of work. Why? Because it is so gloriously green: Indigenous knowledge is recycled and transformed into a comfy, hip, yummy set of food choices. Rather than succumbing to the temptation to invent faddy food rules, Daphne Miller tells purposeful, engaging stories. The writing is slippery and easy to swallow. The message is practical, palatable, and pleasing."

—Harriet Beinfield, coauthor of
*Between Heaven and Earth: A Guide to Chinese Medicine*

"Dr. Daphne Miller presents us with a unique travelogue of healthy eating and in doing so demonstrates that there is much to be learned from the traditional eating practices of the world's premodern cultures. In *The Jungle Effect*, she explores those special people and places on the globe where the diseases plaguing modern society have, by virtue of healthy eating, not appeared. If food preparation is one of the great cultural achievements, then healthy eating is both our birthright as well as our obligation."

—Gail Altschuler, MD, medical director,
the Altschuler Clinic, A Center for Weight Loss & Wellness

# The Jungle Effect

# HARPER

NEW YORK ▪ LONDON ▪ TORONTO ▪ SYDNEY

# The Jungle Effect

*The Healthiest Diets from Around the World—*
*Why They Work and How to Make Them Work for You*

## Daphne Miller, MD

WITH NUTRITION CONSULTATION FROM

### Allison Sarubin Fragakis, MS, RD

All the stories featured in the following chapters are based on actual patients in my medical practice. However, names and many other details have been changed to preserve their anonymity.

THE JUNGLE EFFECT. Copyright © 2008 by Daphne Miller, MD. All rights reserved. Printed in the United States of America. No part of this book may be used or reproduced in any manner whatsoever without written permission except in the case of brief quotations embodied in critical articles and reviews. For information, address HarperCollins Pub-lishers, 10 East 53rd Street, New York, NY 10022.

Foreword © Andrew Weil, MD

HarperCollins books may be purchased for educational, business, or sales promotional use. For information, please write: Special Markets Department, HarperCollins Publish-ers, 10 East 53rd Street, New York, NY 10022.

First Harper Paperback edition published 2009.

Designed by Jennifer Ann Daddio

Library of Congress Cataloging-in-Publication Data is available upon request.

ISBN 978-0-06-088623-3 (pbk.)

12  13  ID3/RRD  10  9  8

To my ancestors—and yours as well.
Their wisdom might just end
up being our saving grace.

# Contents

# Three

# Foreword

BY ANDREW WEIL, MD

With so much confusion and contradictory information about diet and health, where can we turn for guidance? One possibility is to look to the time-tested wisdom of traditional peoples whose freedom from many of the diseases that plague us seems to have a lot to do with their eating habits. That is just what the author of this book, Daphne Miller, MD, has done. She has identified "cold spots"—areas of strikingly low incidence of cardiovascular disease, colon cancer, depression, and other common Western ailments—has visited them, and has studied eating patterns that likely explain the good health of their populations.

Some of Dr. Miller's research will really surprise you. Who would have thought that Iceland was one of the world's cold spots for depression? Iceland is certainly cold, but it's also very dark and gloomy for much of the year. Yet the prevalence of depression of all types, including seasonal affective disorder, is so low that it demands explanation. Icelanders, it turns out, consume more omega-3 fatty acids than any other people, not only from fish but also from roasted seabirds that eat fish, and from the meat and milk of animals that graze on omega-3-rich mosses and lichens. Scientific evidence for the mood-stabilizing and antidepressant effects of these essential fats is solid and growing.

When one observes an unusually high or low incidence of a disease in a population, it is reasonable to ask whether it is due more to genetics or to lifestyle. In this particular case, we have a good idea, because a lot of the Icelanders who emigrated to eastern Canada in the late 1800s were unable to maintain their high omega-3 intake, and became much more susceptible to depression. Because traditional diets are disappearing rapidly as Western processed food and fast food spreads around the world (with chronic diseases close behind), it is possible to draw similar conclusions about the relative importance of lifestyle versus genes in the other disease cold spots described in these pages. In short: lifestyle in general and diet in particular are very important.

Daphne Miller is a physician who studied integrative medicine with me at the University of Arizona. Like me she has a keen interest in nutrition and health and in the practice of lifestyle medicine. She wants to distill the dietary wisdom of traditional peoples and adapt it to our modern world. What is remarkable in her investigations is the uniformity of general principles of the native diets she has studied. Whether in Iceland, Okinawa, West Africa, or the remote canyon country of northwest Mexico, the rules are clear: avoid refined, processed, and manufactured foods; eat good fats and avoid bad fats; eat slow-digesting as opposed to quick-digesting carbohydrates; eat plenty of fresh fruits and vegetables; enjoy a variety of health-protective spices, condiments, and beverages; and get lots of regular physical activity. This is native dietary wisdom in a nutshell, and we can easily adapt it to our own tables.

Dr. Miller helps us incorporate this wisdom by providing recipes from native cultures reworked for ease of preparation in a typical North American kitchen. I can't wait to try the West African *ndole* stew she loves and to make my own version of *goya champuru*, a stir-fry dish containing tofu and bitter melon that I enjoyed in market restaurants in Okinawa while studying healthy aging there. And I will certainly refer patients to Dr. Miller. Her accounts of dramatic clinical success from teaching patients to prepare and eat native foods are inspiring.

This is a groundbreaking book, based on original research, that describes novel dietary strategies for reversing the progression of chronic diseases and maintaining optimum health. Moreover, the native dietary wisdom that

Daphne Miller presents is fully consistent with the findings of cutting-edge nutritional science. I think you will learn as much from it as I did and be inspired to put that knowledge to use.

Tucson, Arizona
October 2007

One

# Introduction

If a diet book is a book that proposes a healthier way to eat, then this is a diet book. And yet, this is not a diet book in the traditional sense of the word. Browse through the health and nutrition section of any bookstore, and you will find that most diet books, whether they are intended to help you lose weight, transition through menopause, or fight heart disease, have one thing in common: they have eating plans that have been invented at a recent point in time by one or a few individuals. This book is quite different. The foods and eating plans recommended are not based on the ideas or observations of a doctor, chef, nutritionist, or supermodel. Nor are they based on laboratory experiments with humans or rats. Rather they have been developed over centuries by indigenous people living on the land in remote places around the globe.

Mind you, the inspiration for this book did not start with a specific intention to write about indigenous diets. Nor did it start with a fascination with isolated cultures, an urge to travel, or a conviction that "natural" or "native" is necessarily healthier. Instead, it started in my medical office in San Francisco in response to seeing so many frustrated patients: men and women who were desperately trying to change their diet to lose weight or deal with a chronic health problem. Most of them had tried dozens of fad

3

diets but had not discovered a healthy way to eat that they could stick to for the long run. For years, I felt that I had little to offer them. But then, as luck would have it, I met Angela. If not for Angela, I might never have thought about the importance of *cold spots* or the healing value of indigenous diets. I certainly would not have set out on the around-the-globe nutrition adventure that has culminated in this book.

Angela was born in Rio de Janeiro, Brazil, the eldest of five children. Her mother was originally from Italy but had grown up in Rio, while Angela's father, whose ancestors were Maués Indians, had been raised in a tiny village on the banks of the Amazon River. As far back as Angela can remember, she was overweight. Hence the nickname *porquinho*, or piglet, that was given to her by her cousins and classmates. She blamed her girth on her mother, who loved to cook heavy, multicourse meals with pasta, all kinds of cheeses, stewed meats, and always a big pastry for dessert. Her father, a physical education teacher, would encourage Angela to play sports, but she usually felt tired and found excuses not to participate.

When Angela was eight, her parents separated. Her father moved to the United States, and her mother, overwhelmed by the five children under her care, decided to send Angela to live with her father's relatives in the rainforest. At first Angela recalled feeling isolated and lonely in her new home. She was depressed about her parents' divorce, hated being sent away, and missed the city and her siblings. She longed for her mother's rich cooking and resented the fact that fruit was considered a dessert and breakfasts featured fish soups, taro, and beans rather than sweet rolls. Then, as the months passed, Angela started to become accustomed to life in the jungle. She recalls many happy moments playing tag with her cousins along the river's edge. As she describes it, "I started to enjoy the food and discovered I really loved living there. Slowly, I began to feel like I had more energy. One day I saw myself in the mirror and was shocked. I looked so good. I looked like a normal-sized child!"

After elementary school, Angela returned to Rio to live with her maternal grandmother. Within months of being back in the big city, she was once again overweight and had lost the sense of vitality she had experienced in the rainforest. By the time she completed high school, Angela felt that she needed a change, so she decided to join her father in New York

City and go to college. Unfortunately, in New York, she continued to gain weight and become increasingly fatigued. By the time she moved to San Francisco in her mid-thirties, she had been on countless diets and had seen dozens of doctors to address her weight and her low energy. She tried thyroid replacement and a wide variety of nutritional therapies, but each of them had only a small, temporary effect. Angela made her first appointment with me because her weight gain had begun to take its toll on her health. Her blood pressure was borderline high, and she was experiencing severe pain in both her knees from her extra weight and inactivity.

During her first visit to my office, Angela told me about her most recent trip to Brazil. Her favorite Amazonian uncle had died at the ripe old age of ninety-six. Rather than simply fly back for the funeral, she had decided to take a leave of absence from her job to spend six weeks in her father's native village. During this visit, Angela noticed some surprising changes in her own body. Despite the fact that she seemed to exercise the same and eat more than she did in San Francisco, she magically dropped fifteen pounds! In addition, her energy level greatly improved, her knee pain got better, and her blood pressure returned to normal. She felt better than she had in years. In fact, the last time that she had felt this good was during her extended stay in the jungle some twenty years earlier. Sadly, within weeks of returning to San Francisco, all these improvements began to disappear. First she felt her energy level sink, then the weight returned, and finally, the blood pressure rose and the knee pain came back with a vengeance. What could explain these dramatic shifts? Angela was in my office to figure out how she could regain her vitality and her Amazonian figure.

After hearing Angela's story, I drew the obvious conclusions. In the rainforest, she must have been eating fewer calories or burning more. If nothing else, she was happier and therefore not using food to treat her stress. I tried to help her recreate some of the foods that she had eaten in Brazil, but this proved to be challenging. Angela was not much of a cook and had trouble recalling many of the ingredients. In the end, I gave her a generic list of low-calorie foods that she might want to consider, and we talked some more about exercise.

Honestly, I probably would never have given this exchange a second thought if not for my own experience in the Amazon rainforest six months later. My husband and I are always looking for places to escape a freezing San Francisco summer. This particular year, we decided to spend a month volunteering in a tiny Peruvian village deep in the Amazonian basin. By sheer coincidence, the village, called Las Palmeras, happened to be less than a couple days' slow boat ride west of Angela's family homestead. Las Palmeras, accessible only by boat, was nothing more than a handful of sugar cane fields, a small cinderblock school, a clinic, and a scattering of open-walled thatched huts on stilts. Each hut housed an extended family along with a couple of dogs and, more often than not, a pet monkey. The village had no electricity or running water, and the villagers relied on the river at their doorstep for all their plumbing needs. The only forms of commerce in the village were two tiny rum distilleries that still used a horse-powered press to squeeze the sugar cane. There were no markets, and everyone seemed to eat what they grew and nothing more. Across the tributary from the village was a basic bamboo-and-thatch explorer's eco-lodge built by an ex–Peace Corps volunteer. This is where we stayed and ate many of our meals in exchange for our services.

While my husband and kids volunteered in an after-school reading program, I spent my mornings working in the clinic. The clinic had been founded well over a decade earlier by Linnea Smith, an intrepid Wisconsin internist who was affectionately known to the locals as "La Doctora." This tin-roofed structure, built on stilts and relying on solar electricity, was the only health-care facility for miles and miles. All things considered, it did a remarkable job of providing basic health care to the locals who would arrive on foot and by dugout canoe, sometimes traveling days from their little villages scattered along the Amazon's many tributaries.

During my month in Las Palmeras, I made some interesting observations. First of all, while working in the clinic, I saw my share of snake bites, machete wounds, tetanus, malaria, and severe infant diarrhea: all health problems that could be easily prevented or treated with modern innovations such as indoor plumbing, vaccines, and antibiotics. However, I did not encounter a single person (other than the tourists) who was struggling with the diseases that I see on a daily basis in San Francisco: obesity, heart disease, diabetes, and depression. Even among the weathered elderly, the main ailments seemed to be infection and injury.

Second, outside of the clinic, I noticed that a transformation had taken place both in the Peruvian tour guides and the North American volunteers. While most of the tour guides who escort foreign travelers to the jungle lodge were originally from the village, they now spent the bulk of their time in the city of Iquitos. There they do relatively little physical activity and eat mainly a Western diet consisting of pizza, pasta, and sodas. Their relatives in the village of Las Palmeras, however, have continued their traditional lifestyle. They labor in the fields and eat food that is raised and prepared locally. As my family made friends with these guides and met their families in the village, we were shocked by the fact that the guides looked completely different from everyone else in their clan. The guides sported jowls and pendulous pot bellies while their village brothers were fit and sinewy. By contrast, several of the long-term volunteers from the United States noticed the opposite changes in themselves. They were pudgy and out of shape upon arrival in Peru and were amazed to see their bodies trim down over the following weeks in the jungle. Despite the high humidity and the blazing heat, the volunteers told me that they felt more energized than they had in a long time.

Noticing these changes, I suddenly recalled Angela's story. I started to pay closer attention to what could be causing this jungle effect. Why was it that villagers and volunteers alike seemed to be fit and slim while the guides were falling apart? The nearest road or car was a long boat ride away at the river's mouth, so our only options for transport were our own two feet or a dugout. Nonetheless, I did not feel that this could account for the dramatic change since the heat kept us moving slowly and all the North Americans in our group exercised routinely even before they came to the Amazon.

I decided that the magic I was seeing must be in the food. Angela had commented that she felt like she ate more in the jungle, and I had exactly the same feeling. Therefore, the answer did not seem to have anything to do with food *quantity*. I began to look more closely at the *quality* of our jungle diet. While the occasional motor boat would bring in supplies from the nearest city, Iquitos, much of what we ate at the lodge came from the fields that line the river's banks or from the river itself. As a result, we were eating very differently than we did in our usual lives. Breakfasts included fruit, pan-cooked fish, or a freshly-laid scrambled egg. For lunch there was more fresh-caught fish, plantains, red hot pepper salsa, and piles of something that the locals called "jungle spaghetti"—shredded hearts of palm. Dinners consisted of fish or free-range chicken, boiled or baked yucca, and more hearts of palm. Fruits with names I could barely pronounce were the main dessert. Every front porch had a branch of fresh bananas. On the rare occasion when hunger reared its head in between meals, we would simply break one off for a filling snack.

It occurred to me that, unlike so many fad diets that were based on one person's understanding of healthy eating at that moment in time, what I was eating in the Amazon was a tried-and-true diet—a diet that had evolved over generations and had helped villagers maintain their fitness and stave off modern diseases for centuries. To be sure, the magic of the jungle meals was partly in the ingredients. However, after spending days writing down lists of foods, I realized that the healing nature of these foods lay mainly in their proportions, combinations, and sequence in the day. In other words, a list of ingredients would serve little purpose without the actual recipes.

Since cooking seemed to be under the purview of women, I began talking to the nurses at the clinic and the women in the village. The male cooks at the lodge also gave me some ideas. As best as I could, I tried to collect cooking methods and cooking times. Of course, some of these foods could not be found outside the jungle, but many were available in standard North American supermarkets. In the end, I had a dozen recipes neatly written in my journal that I brought back as precious souvenirs from my travels.

When I e-mailed the recipes to Angela, she called me back immediately. She was surprised and amused at my unconventional approach. Never before had a doctor actually prescribed her a set of recipes. Not being much of a cook, she told me she would give them her best try. A month later she called me back delighted. She told me that she had gotten some of her jungle energy back and was really enjoying the food. Two months later, I saw her in my office. She was exercising regularly and cooking the recipes I had given her, and she had incorporated some new ones that her aunt had sent her. Her weight was down almost fourteen pounds, and her diet felt easy to maintain. She said that a number of her friends had commented on her weight loss and asked her what she had done to achieve this. Without hesitation, she had answered, "I put myself on the Jungle Diet."

# 1. Dining in the Cold Spots

Like Angela, most of the patients that I see daily in my medical practice are trying to prevent or treat a chronic health problem, lose weight, and preserve vitality. For many years, I felt ill equipped to help them achieve their goals. After four years of medical school, three years of residency training, and two years in a postgraduate fellowship, this is a hard confession to make.

Of course, I was well versed in using medications. Initially I found it satisfying to watch how rapidly many of these drugs took effect—sometimes lowering blood pressure, blood sugar, or cholesterol levels in a matter of days. However, after several months of practicing medicine out in the real world, it became obvious to me (and certainly obvious to my patients) that there were many unintended side effects from the treatments that I was dispensing daily with my prescription pad. Furthermore, it seemed that my standard approach was not getting at the root of so many of these health problems: the foods my patients ate on a daily basis.

These days, the majority of serious health problems that we are experiencing in the United States can be traced back to a poor diet. As a matter of fact, unhealthy eating habits, lack of exercise, and smoking are now the

9

three leading causes of chronic disease and death in most parts of the industrialized world. It has been estimated that at least 80 percent of medical cases involving heart disease, stroke, and type 2 diabetes and 40 percent of cancer cases could be avoided by following a healthy diet, participating regularly in physical activity, and avoiding tobacco. More troubling is the fact that these modern health problems are showing up in young people. In my practice, I have started seeing teenagers with early signs of type 2 diabetes and heart disease!

## What Are Some Modern Chronic Diseases?

Modern chronic diseases are long-term health problems that are relatively rare in traditional (or indigenous) societies but increase in parallel with the degree of industrialization in a given area.

Examples include:
- Stroke
- Diabetes
- Heart disease and hypertension
- Cancer
- Emphysema
- Asthma
- Depression
- Dementia
- Gout
- Inflammatory bowel disease
- Autoimmune diseases, such as lupus and rheumatoid arthritis

Given this information it seems logical to me that the foods we eat (along with exercise and giving up tobacco) should be the first line of defense against modern chronic diseases. Food is a powerful medicine. After all, it is the medicine that most of us take willingly at least three times a day without skipping a dose. But here lies the big challenge: while most of us can agree that food is a very important therapy, how do we use it to prevent or treat chronic disease? *Which* foods are actually the healing ones? And, most importantly, how do we make these dietary changes in our daily lives?

In an attempt to help my patients, I set out to learn more about nutrition. I went to conferences and pursued training programs taught by internationally recognized nutritionists and physicians. Using what I learned, I started to counsel my patients on how to make dietary changes: I would launch into a discussion of healthy diets and, of course, exercise. I would rattle off the standard recommendations from groups like the American Heart Association (AHA) or the American Diabetes Association (ADA). "Decrease total daily kilo calories to 2,000, keep saturated fat intake below 2 percent of calories, cholesterol intake should be no more than 150 grams per day, salt may be a factor so limit daily sodium to 2 grams daily, stick to 2 servings of refined starches. . . ."

Unfortunately, this was not the best approach. Few of us spend our days calculating our food intake in percentages and serving weights. Therefore, the general AHA or ADA recommendations have little bearing on our daily reality. Even if we are able to follow these recommendations, I am not sure how far they get us. Let me explain.

## Nutrient Splitting and Wishy-Washy Nutrition Research

Much of modern nutrition science is based on the idea that there is a perfect amount of macro nutrients (proteins, carbohydrates, fats) and micro nutrients (iron, calcium, folate, etc.) that will help to preserve health or prevent a specific disease. Therefore, when counseling patients, the goal is to try to give them a prescription for how much of each of these nutrients they should consume each day. This practice of breaking nutrition down into its constituent parts is sometimes referred to (somewhat pejoratively) as "nutrient splitting."

But how do we arrive at the perfect nutrient split? How do we know the ideal amount and proportion of these recommended nutrients? Usually this is dictated by the research du jour and the latest studies to roll off the academic presses. I am sorry to report, however, that this research is often quite flawed—not through any fault of the researchers but simply because it is a nearly impossible task to perform a "real-life" nutrition experiment on a human.

First of all, many foods take years to have a real effect. And yet, the longest that most humans will subject themselves to a controlled laboratory rat-style food study is a couple of months. For example, let's look at the research on soy foods and breast cancer. There is evidence to suggest that adolescence may be the critical time for eating whole soy foods such as tempeh and soy beans to prevent breast cancer from developing forty-five to sixty years later. However, most of the published soy studies have focused on giving high doses of soy foods or soy extracts to mature women for several weeks to months. Given that the lag time between exposure (soy) and outcome (breast cancer) is likely to be decades, it is understandable that these studies have not netted very much useful information. In fact, some research has suggested that high doses of soy this late in life might even be harmful.

Next there is the problem of comparing carrots to carrots. For example, a carrot grown organically in your back yard can be vastly different from another that is grown commercially on a large farm. The nutrients and chemicals in the soil, the weather, and the type of seed that the carrot originated from will all change the way that it behaves in your body. How the carrot is prepared and eaten are also important considerations. Factors such as whether it is raw, cooked, boiled, or simply the beta-carotene from a carrot reduced to a pill form can cause this same food to have very different effects. A recent illustration of this phenomenon is the ATBC (alpha-tocopherol beta-carotene) study. Previously, a whole host of studies had suggested that people who consumed foods rich in vitamin E and beta-carotene had lower rates of cancer. However, when smokers were given a pill containing a high dose of these same vitamins, there was a paradoxical effect: the supplement actually *increased* their likelihood of developing lung cancer and heart disease! As we see in this example, even if vitamins are *extracted* from whole foods, they can behave differently in their extracted form than they do when they are consumed in a real piece of food.

The *interaction* between foods and nutrients in foods poses yet a third big challenge to studying nutrition. Many of us know that vinegar or citrus helps release the iron in spinach. But there are hundreds of these interactions that take place behind the scenes with each meal. Most food scientists agree that these interactions are much too numerous and complex for us to begin to account for all of them in a scientific setting.

Finally, the physical health effects of food cannot be separated from the expectations we have about that food. In one interesting experiment, researchers showed that our blood sugar levels can vary depending on our *perception* of sweetness in a given food. This can happen independent of how much sugar is actually found in that food. By this same token, it has been shown that believing whether a food is healing or harmful can also play a role in how it affects our body.

With all these details to consider, you can see why studying the effects of food in our bodies often leads to confusing and contradictory results. It leaves doctors like me giving recommendations that are all over the map. One year I am telling patients to consume less salt, but then the next year I change my message: salt might be okay. Cholesterol in food raises your cholesterol, or wait a second, maybe not. Don't eat animal products. Maybe animal products but not red meat. Margarine is better than butter. Oh no! Margarine has trans fats. Maybe you should go back to butter. Eat more folate rich foods; they prevent heart disease. No, not heart disease, but maybe cancer. High fiber does not prevent colon cancer; actually, yes, it does. No wonder patients became frustrated by this wishy-washy advice.

In search of a healthier way to eat, some of my patients also consulted nutritionists or bought books with specific diet plans. Often they joined weight-loss centers and began eating prepackaged weight-loss meals. Sometimes, at the follow up appointment, we would rejoice that there were pounds lost and improved laboratory values such as cholesterol and blood sugar levels. But these changes never seemed to last very long. For most people, the prescribed diets became bland, boring, or simply too difficult to follow in the long run. Many patients found themselves going back to their old ways of eating, which usually featured frequent snacking, fast foods, and takeout or precooked frozen meals. Time and time again, I would see the pounds creep back and the laboratory numbers slowly return to their starting point. Usually after a yo-yo through several cycles of weight loss and weight gain, we would finally agree to start medications. Even with medication, patients' health problems would slowly get worse. Obviously, something was not working. It was time for a new approach.

## Traditional Diets: A Healing Secret on the Verge of Extinction

My experience with Angela gave me an idea. Could my patients and others benefit from learning about traditional diets around the globe? After all, Angela's jungle diet had stood the test of many prior generations, and I had seen living proof of its benefits. It reflected an accumulated knowledge about foods that had been passed down through many generations in a natural environment where people needed to be fit to survive. We did not have to worry about the blandness or boredom of one-size-fits-all diets —these were recipes that had been refined for centuries and were rich in flavor and tradition. Furthermore, we did not need to worry about nutrient splitting and about which specific ingredients were lowering cholesterol or blood sugar or preventing heart disease. We could just assume that the traditional recipes contained the healthy proportions of the foods and that the actual medicine was in the interplay of all the foods.

Of course, I am not the first person to have this idea. As a matter of fact, nearly eight decades before my trip, a dentist named Weston Price had made similar observations after spending time in the Peruvian rainforest. Dr. Price's main interest was the connection between diet, oral health, and tooth decay. As he traveled to remote societies around the globe, from the Peruvian Amazon to the Outer Hebrides to Central Africa to the islands of the South Pacific, he carefully noted the state of people's teeth and jaws and the quality of their diet. In his book, *Nutrition and Physical Degeneration*, he noted that people living traditional lives as hunter-fisher-gatherers were practically cavity free while people living modern lives and eating preserved and processed foods had experienced an explosion in the number of dental caries. They also seemed to be suffering from other chronic diseases such as diabetes, heart disease, arthritis, and cancer. In fact, tooth decay seemed to be the dental version of a modern disease. He concluded that these indigenous foods held special healing qualities and became a proponent of abandoning our processed Western meals in favor of more traditional diets.

Of course, the world has changed dramatically since Dr. Price made his voyage. Every corner of the globe has been touched in some way by modernization and mechanization. Thanks to powerful transnational corporations and a global economy, the so called "McDonaldization" is nearly complete; even the poorest and most remote communities have access to packaged foods that are laden with processed fats, refined sugars, and man-made chemicals. I personally have hiked to the deepest recess of the Cameroonian rainforest and, at journey's end, been offered a warm Coca-Cola by a Babinga pygmie tribesman. The average Mexican now drinks 487 cans of Coca-Cola per year, twice the amount consumed ten years ago. In the same period of time, India has gone from being one of the lowest to one of the highest consumers worldwide of refined vegetable cooking oil. As a result, it is no longer possible to find an indigenous or native diet in its purest form. Even in the rainforest in Peru, the medical assistant at the clinic was starting to extract rotten teeth on a weekly basis, whereas the previous generation had no need for this procedure.

While authentic indigenous diets may be slowly disappearing, there are still some isolated communities around the globe who remain relatively free of modern diseases. Like Weston Price, and other nutrition explorers who came after him, I believe that these communities have much to teach us. But where are these communities? How do we identify them, and how do we go about learning from them?

## Hot Spots and *Cold Spots*

The first place I began to look for answers to these questions was in my enormous collection of cookbooks from around the globe. (Some people collect trinkets or T-shirts when they travel; I collect cookbooks.) I thought I would use the recipes as a jumping off point for learning more about a specific community. My search uncovered many interesting recipes considered to prevent or treat a range of problems from gout to cancer. However, I am always wary of health claims founded on lore or hearsay, and I wondered if there was any real evidence to support the supposed health benefits of these meals. On a quest for proof, I began to interview my colleagues in anthropology and medical anthropology who were interested in indigenous food. They were knowledgeable about the eating patterns of different cultures from around the world. However, because they were not trained in medicine or nutrition, they were not able to help me understand which diets were especially healthy or why certain foods had a strong healing quality. Then I asked some savvy nutritionists. They knew a lot about nutrients but had very little to say about indigenous diets. Finally, I decided to switch gears and take a look at the epidemiology literature. What I found was intriguing.

Epidemiology is the study of disease trends in populations. Epidemiologists are like detectives, except that the perpetrator they are trying to track down is a disease or a risk factor rather than a person. Epidemiologists sometimes refer to certain places as disease *hot spots*. Hot spots are places where there are unusually high numbers of people suffering from a particular disease. For example, for reasons yet unknown, the San Francisco Bay Area happens to be a hot spot for breast cancer. In general, epidemiologists get very excited about hot spots because they are places where you find many cases of a disease and can really study it in detail. They flock to hot spots to interview people, take blood samples, and do any other investigative work that will help them uncover the true causes of a disease.

If there are hot spots, then it stands to reason that there should also be *cold spots*. These would be places where surprisingly few cases of a disease are found. Epidemiologists who are doing cold spot studies are trying to figure out why a certain disease is rarely seen in a given population. Digging through the medical literature, I discovered some discrete cold spots scattered around the globe. These were places that not only had lower rates of chronic disease in general but also had especially low rates of one specific disease such as heart disease, diabetes, or breast cancer. As one might expect, they were also places where the indigenous diet was alive and well. At last, I knew what I needed to do. I had to focus my attention on the traditional foods in these cold spots. By taking this approach, I would discover ingredients and recipes that truly maintained health and prevented or treated chronic disease.

## Hot Spots and Cold Spots

- Hot spot—Place or community where there is an unusually high number of people suffering from a particular disease.
- Cold spot—Place or community where there is an unusually low number of people suffering from a particular disease.

Part Two of this book features the cold spot diets for the following modern chronic diseases: type 2 diabetes, heart disease, depression, colon cancers and other bowel problems, and, finally, breast and prostate cancer.

In the remainder of Part One, we learn about the basic anatomy or *key healing components* of these cold spot diets and we find out why so many of us have lost these key elements in our modern way of eating. In this section, we also explore the relative importance of taste buds, genes, and personal health history in determining which indigenous diet(s) to incorporate into our daily cooking and eating experience. In Part Two, we begin our travels around the globe. Each travel chapter begins in my medical office and is inspired by a patient with a set of health concerns. This patient launches us on a journey to a remote cold spot, a place where there are very low rates of their particular health problem. During each of these voyages, we discover how the local foods play an important role in preventing or treating specific chronic diseases. We also uncover nutrition tips that can easily be integrated into our own lives.

The cold spots in this book vary wildly in their locale and topography, from seaside to deep canyon, from the tropics to the desert, and all of them are in relatively remote places. In most cases, people living in these zones have actively resisted pressures to modernize or assimilate and are often wary of intruders. While this is certainly understandable, it has sometimes made it difficult for me as a visitor to glean much useful information. Therefore, while I have traveled to all of these places, I have also relied on a wealth of knowledgeable sources, be they community members, natives now living abroad, indigenous healers, local physicians, medical researchers, anthropologists, psychologists, adventurers, or chefs.

## The Migration Effect and the Benefits of Cold Spot Eating

Each patient whose story is featured in this book has a different ethnic and cultural heritage as well as a distinct set of health issues. However, all of them share one thing in common: they belong to one of the first generations in their family to develop their chronic health problem. Why is this the case? Like so many people across the globe, these patients have experienced a phenomenon which social scientists call the "migration effect." The migration effect is essentially the opposite of the Jungle Effect. If the Jungle Effect describes the health benefits associated with traveling back (either literally or virtually) to a more indigenous way of living, then the migration effect, by contrast, describes the modern chronic health problems we face when we abandon an indigenous lifestyle in favor of a more industrialized one. While the migration effect is experienced by people who have emigrated across continents and oceans, it also afflicts those whose ancestors have not budged more than several miles in the past thousand years. For example, the Pima Indians of North America, Native Hawaiians, and Australian Aborigine experienced a similar downward spiral in their health when modernization was brought to their very own doorsteps. There are many factors that may contribute to the migration effect though most of the research indicates that eating more processed foods and adopting a sedentary lifestyle are the main culprits.

Even though most of the patients you will meet in the following stories are struggling with serious health issues, all of them report positive results from changing their diet. As their stories illustrate, once they learned about cold spots and indigenous diets, they began to change the foods that they bought and cooked. A return to an indigenous style of eating helped them accomplish a diverse set of health goals including weight control, blood sugar and cholesterol management, and improvement in energy and mood. Moreover, most of them report that they have finally found a style of eating that they can stick with for the long run. My hope is that you will experience the same success as you bring cold spot cooking into your own home.

## Cold Spot Cooking in Your Own Kitchen: Small Effort, Big Payoff

I would love to be able to offer up some quick fixes or tell you that there is a wonderful selection of prefab dinners or restaurant meals that will give you the health benefits described in this book. Alas, it is not so. To truly experience the positive effects discussed in the following pages, you will need to start cooking. Therefore, while I hope that you thoroughly enjoy the nutrition adventures and expert advice offered in the following pages, my ultimate goal is that this book, food stained and water logged, finds a home on your kitchen counter as a favorite cookbook.

Yes, cooking from scratch takes more time and effort than going to a fast-food drive-through or a standard restaurant. It even takes more time than pulling the lid off of a frozen dinner and popping it in the microwave. However, considering the huge health benefits, it does not take that much more time. Most of you work long hours and have limited time to shop or get meals on the table to satisfy hungry families. To meet your needs, I have tried to make things as simple as possible. In the end, I have assembled a set of authentic recipes that are relatively fast and easy to make. Each one has been tested on dozens of people, including friends and patients, to ensure that it fits this bill. Most of the ingredients are found in standard supermarkets and are easily interchangeable depending on what is available and in season.

## Be Wary of Imposters

While some ethnic food restaurants try to use traditional ingredients, this is often not the case. Sometimes the food is an entirely new invention made for North American palates, while other times there are ingredient substitutions that turn a healthy dish into an unhealthy one. Recently, as I was waiting to use the restroom in my favorite Thai restaurant in San Francisco, I was reminded of this phenomenon. Tucked back in the kitchen, I spied a huge case of Skippy crunchy peanut butter. So this was the secret ingredient in all those peanut dishes! Thai recipes that were originally meant to include whole peanuts, rich in monounsaturated fats and protein, now contained Skippy, a nut butter mixed with other manipulated fats, that happens to have sugar as its second ingredient. This illustrates how important it is to cook your own foods and do most of your cold spot eating at home.

You will see that there are certain healing ingredients that pop up in all the diets, and therefore, each chapter builds nicely on the next. Some of these ingredients might be well known to you, but you will learn new ways to use them. Others are familiar, but for whatever reason, you do not use them on a regular basis. And still others are right within your grasp, but you have yet to discover them. I assure you that once you get to know the recipes and become comfortable with the ingredients and the variations, each meal will become even easier to make.

While some people may prefer to focus on one specific cold spot region due to their particular health concerns or their appreciation for the tastes from that area, others will want to rotate through the whole book. Each chapter has at least two recipes for breakfast, lunch, and dinner. In total, you have fourteen full days of eating (plus some special feasts) if you cook from the whole book. This number was not arrived at by accident. After many interviews, I have discovered that most people who cook at home have about fourteen meals that they make on a regular basis. Keeping a dozen-plus meals in my repertoire not only guarantees a manageable shopping list but also ensures enough variety that no one in my family gets bored.

## The Indigenous Diet Score

Throughout this book, I have made every attempt to focus on whole foods and avoid nutrient splitting. However, many of my patients have asked me for some additional guidelines when planning their week's worth of eating. For this reason, I have worked with my friend Allison Fragakis, who is a brilliant nutritionist and a gifted cook, to give each recipe an *indigenous diet score*. This can help ensure that you are getting an optimal array of foods over the course of your day and week.

Aside from the dramatic health payoffs, there are other potential upsides to spending a bit more time preparing your food. I find cooking to be a relaxing (almost meditative) experience that helps me decompress from a busy, demanding day. The zen of chopping should not be underestimated, especially when I have a good, sharp cooking knife. I also find that if I give other family members or guests specific kitchen chores (washing the greens, peeling the onions, setting the table), it is a wonderful way to spend time together. For example, my son Emet has become an expert tortilla maker and no one dares challenge his turf. Finally, there is the exercise that I get in my kitchen from moving from fridge to counter to stove, all the while peeling, chopping, and mixing. Certainly more calories are burned there than from behind the wheel of my car tooling up to the fast-food drive-through window.

All this being said, cold spot cooking does take some time from your day—time that might seem hard to justify in a typically busy week. Whenever I find myself grumbling about those additional thirty to forty minutes, I find it helpful to recall the words of my colleague who is a kidney dialysis specialist at the local hospital. Many of his patients have end-stage kidney failure caused by their hypertension and type 2 diabetes. "You know, Daphne," he said with a sigh, "it always amazes me how people can find the hours in their day to sit hooked up to these dialysis machines. Yet, before they got sick, many of them could not find a fraction of that time to eat well and exercise so that they would not end up here in the first place."

# 2. Anatomy of an Indigenous Diet

For several thousand years, people living in Central and South America have used a process called *nixtamalización* to treat their maize. This entails boiling the corn in an alkali solution containing lime and water. While this is a handy way to remove the tough outer coating of the corn kernels, there is also a major health benefit to using limewater: it enriches the corn with essential nutrients such as calcium, niacin, and certain amino acids. Without nixtamilization, Mesoamericans would have suffered from severe vitamin deficiencies, including niacin deficiency (also called pellagra).

As I began to learn more about the diets from various cold spots, I discovered dozens and dozens of similar preparation techniques and food combinations that further enriched the nutritional value of traditional meals. It made me wonder how people living so many years ago could invent diets that were full of culinary wisdom and could pass the muster of even the most critical modern-day nutrition experts. After all, there is nothing to suggest that our ancestors had a formal understanding of the nutrition concepts that are now considered common knowledge. Take the Tarahumara of Northern Mexico for example. They are a tribe that has used nixtamalization for thousands of years. Surely their predecessors could not have known about a vitamin called niacin or a disease called pellagra. While some ancient cultures were quite sophisticated with written languages, astronomy, and even mathematics, I am certain that none were aware of the fact that there is a group of fats with an omega-3 carbon structure or a collection of micronutrients that we refer to as *antioxidants*. Furthermore, there is little evidence that these cultures had an in-depth understanding of the chronic diseases that their diets seem to be so effective in preventing. So, was it all just dumb luck or was there something more conscious and purposeful to their food selection?

## What Makes Indigenous Diets So Healthy?

I called Harriet Kuhnlein, PhD, to help me better understand how isolated communities around the globe could invent such healthy diets. Kuhnlein is an anthropological nutritionist by training and the director of the Center for Indigenous People's Nutrition and Environment at McGill University in Canada. She explained to me that *indigenous diets are born when a group of people use their traditional knowledge to make a complete diet using local foods*. These diets have slowly evolved in a natural setting and have stood the test of hundreds of years. Although Harriet Kuhnlein feels that some indigenous diets offer a more optimal mix of nutrients than others, they are all (almost by definition) nourishing because they rely on *fresh, in-season ingredients*. All the foods in an indigenous diet are either produced locally or purchased (traded) within a limited geographic area.

### Indigenous Diet Key Component #1

Foods that are local, fresh, and in season

This explanation made perfect sense to me, but I wondered how this nutritional knowledge was gathered and how each unique cuisine was created. What led to the development of all the healthy food preparation techniques that I encountered in my travels? How did Tarahumaras figure out how to nixtamalize their corn? What inspired people in Crete and Iceland to ferment their milk into a yogurt rich in immune-enhancing bacteria? How did Cameroonians figure out that soaking their manioc would remove toxic residues? And what gave the Okinawans the idea that slimy seaweed would offer them such a terrific complement of nutrients? To learn more about this, I called a number of anthropologists, sociologists, and ethnobotanists, all of whom gave me very helpful information. However, interestingly enough, it was a food psychologist who finally gave me an answer that made everything fall into place.

## The Three Basic Food-Selection Instincts

University of Pennsylvania psychologist Paul Rozin, PhD, has spent most of his career thinking about food selection and eating behaviors. Over the last five decades, his interests have led him to study how people of different nationalities make their food choices, how small children establish their likes and dislikes, and even how rats are drawn to their favorite edibles. Rozin explained to me that all humans past and present value convenience when it comes to obtaining food and, therefore, tend to select foods that are readily at hand. But how did our ancestors select their specific diet out of an array of available plants and animals? To do this, they called on three basic human instincts which Rozin roughly categorizes as 1) imitative, 2) sensory, and 3) post-ingestive. These instincts are there to help us negotiate a very dangerous situation: the sampling of new foods. Should we suddenly find ourselves marooned on a deserted island, we would rely on these internal guides to help us identify safe food sources and avoid poisonous ones. These are also the factors that guided our ancestors, no matter where they were located around the globe, as they collectively developed an indigenous cuisine.

## Instinct #1: The Power of Imitation

Imitation, according to Paul Rozin, plays a central role in helping us select, prepare, and even eat our foods. Suppose your ancestors, much like the Vikings, happened to row to a faraway island to start a new life. How would they have decided what to eat on the island? If another group of people already existed on the island, your ancestors' first instinct would likely have been to look to these people to get a clue about which foods were edible. If there were no signs of human life, your ancestors might have found themselves imitating the local animal population or trying to recreate their previous diet by selecting foods similar to ones back home. To understand imitative behavior at its most basic, just look at how a toddler selects her food. In general, her choices are inspired less by what she is *told* to eat than what she actually *sees* parents and peers eating. Any experienced parent will tell you that the best way to get a broccoli-hating child to sample this food is to have another child sitting nearby enthusiastically eating broccoli.

In a natural setting where people remain connected to the land and its resources, the imitative instinct allows valuable information about how to gather and prepare food to be passed down generation to generation. If one of your ancestors discovered a plant that kept her feeling vital and strong, then her children would copy her and incorporate the plant into their daily diet as well. As indigenous diets evolve, this imitative information turns into a set of food-related customs. Food customs, be they religious or secular, often play a vital role in cementing a group's identity. They also ensure that any given generation is linked not only to their parents but also to their grandparents and the many generations that came before.

### Indigenous Diet Key Component #2

Food cultivation techniques and recipes passed down through the ages

In most traditional settings, the imitative instinct plays an important role in determining not only *what* people eat but *how* they eat it. Virtually every cold spot culture I discuss in this book places a huge emphasis on communal eating; unless one is miles away from other humans, one simply does not eat alone. In these group settings, the imitative instinct helps us to learn healthy table manners and patterns for eating. For example, in Cameroon and Crete, one is absolutely expected to eat slowly—a practice that greatly improves digestion and, interestingly enough, often results in eating less food overall. In Okinawa, the dictum *Hara Hachi Bu*, or "eat only until you are eight parts full," is a reminder not to overeat. There it is considered much more polite to honor the messages from your stomach than to eat everything on your plate. The periodic Easter time fasting that is practiced in both Crete and Copper Canyon is yet another type of imitative ritual that has health benefits: individuals who observe these forms of modified fasts (either for religious or health purposes) have lower cholesterol levels and body weight even when they are not in a fasting period.

## Indigenous Diet Key Component #3

Food traditions:

- Communal eating
- Eating for satiety rather than fullness
- Observation of fasts and other food rituals

## Instinct #2: The Power of the Sensory

Getting back to the scenario we discussed earlier: what are some of the factors other than imitation that would have guided your ancestors as they made their food choices on their new island?

Not surprisingly, the taste and smell of food would have strongly influenced their dining decisions. Psychologist Paul Rozin explained that there are four basic food experiences that are universally attractive to all humans: sweet, salty, meaty, and fatty. Barring any preconceived notions, such as social taboos or nutrition information, these are the first types of foods that your ancestors would have selected.

At first glance, our instinctive attraction to sweet, salty, meaty, or fatty foods might seem like a curse when it comes to choosing a healthy diet. Indeed, as we will discuss in the next chapter, if faced with abundant quantities of modern processed foods, this instinct is likely to get us into trouble. However, if you understand that indigenous diets evolved in a natural setting, then these food preferences actually make for excellent eating choices. To learn how these taste preferences can nurture us, let's first look at "sweet and salty" and then at "meaty and fatty."

## Sweet and Salty

Aside from the occasional honey stash from a beehive, most sweet things that occur in nature are plants, including fruits, some vegetables, flowers, and grains. Unlike modern sweet foods such as candies and baked goods, most naturally occurring sweet foods contain relatively little sugar (sucrose, galactose, or fructose), and therefore they do not cause increases in blood sugar and insulin levels. (As we will discuss in the chapters that follow, high blood sugar and insulin levels are linked to a whole host of modern chronic diseases.) Sweet foods, as they exist in nature, also happen to be laden with an array of vital nutrients such as enzymes, proteins, healthy fats, minerals, fiber, vitamins, and other antioxidants.

## A Look at Acidic, Spicy, and Bitter

Why are most toddlers disgusted by coffee while many adults are happy to drink it on a daily basis? We intuitively know that acidic, spicy, or bitter foods are an *acquired* taste. But what does this mean? Like our ancestors, many of us have initially steered away from these more challenging tastes since we are hardwired to associate them with poison and spoilage. Sooner or later, however, we are likely to find ourselves imitating an adult who enjoys these foods, and eventually we acquire an appreciation for them as well. In this instance, the imitative instinct permits information about the value of acidic, spicy, and bitter foods to be passed on to the next generation.

Similarly, salty edibles that occur in nature such as fish, sea vegetables, root vegetables, and leafy greens have relatively little sodium and are coupled with many nutritious ingredients. Therefore our attraction to salt makes for healthy food choices when we are faced with a selection of unprocessed foods. Whole foods, whether they are sweet or salty, also contain a fair amount of fiber. Fiber has the unique property of staying in our gut and absorbing water. As a result, it helps us to feel full without taking in an excessive amount of calories. *Therefore although our ancestors may have been lured in by their sweet or salty taste buds, pretty much any food that they would have selected would have given them a sizeable dose of nutrients and filling fiber in the bargain.*

## *Meaty and Fatty*

An attraction to meaty and fatty foods might also seem like a counterproductive human instinct, given that meat-filled and/or high-fat diets have been linked to a number of modern diseases, including breast, prostate, and colon cancer, diabetes, and heart disease. However, as with our attraction to salty and sweet, there are some factors that could make this meat-and-fat-seeking behavior a healthy one in a natural setting where we are relying on nearby food sources.

First and foremost, there is the issue of *quantity*. Humans have always preferred an easy food source, and in most natural settings, putting meat and other animal products on the table is far from easy. It requires that we expend a lot of time, money, or physical energy. Domestic animals, raised for meat or dairy, must graze in fields and pastureland that otherwise could be planted with crops offering a direct source of nourishment for humans. Hunting, on the other hand, requires dexterity, speed, and strength. Regardless of how the meat is obtained, it needs to be slaughtered by hand and cleaned. (After spending some time watching my local butcher at Drewes Brothers Meats carve up a lamb, I now understand that this demands no end of skill.) Similarly, milking a cow, goat, or sheep, (especially without an automatic milker) calls for a fair amount of elbow grease. For all these reasons, most traditional cuisines use animal products sparingly, often reserving them for special occasions and feasts. As you will see in the cold spot recipes that follow, daily meals rely on more readily available sources of protein such as legumes, nuts, grains, and vegetables. In addition, because animal-based foods are so precious, there is no waste. In the case of meat, almost all organs are eaten, and in the case of dairy, the whey from milk is put to many uses. It so happens that organ meats and whey are some of the most nutritious parts of animal food sources and probably play a significant role in making indigenous diets so healthy.

In addition to quantity, the *quality* of indigenous animal products is also a consideration. Almost by virtue of being locally obtained, meat and dairy will be nutrient-rich since the animals will have fed on open pastureland (otherwise known as "free-range") or on wild foods. Eskimo tribes who live near the North Pole offer us an extreme example of how meat quality can make a difference in the overall nutritional value of a diet. Until relatively recently, these tribes spent more than half the year eating little more than locally caught whale meat, seal, fish, sea birds, and caribou. Despite their high protein (50 percent) and fat (30–40 percent) diet, Eskimos who ate this indigenous diet had some of the lowest rates of heart disease and cancer in the world. When nutritionists and epidemiologists studied members of the Eskimo community, they discovered that they had exceptionally high levels of two omega-3 fats in their blood: EPA and DHA. As we shall see, these are fats known to play an important role in the prevention of many chronic diseases. The Eskimos obtained these fats from dining on seafood and wild game. These animals, in turn, synthesized their omega-3s from eating sea greens and vegetation on the tundra.

These same nutrition researchers were surprised to discover that the Eskimos also had adequate amounts of essential nutrients such as B vitamins and vitamin C. These are vitamins that are usually obtained from eating plant foods. On their meat-packed diet, how could the Eskimos possibly have enough of these vital nutrients? It turns out that once again, fish and caribou were the main sources. In the case of fish, these vitamins were obtained by eating other fish and sea vegetables, and the caribou were passing on nutrients from the scrubby lichen and moss that they munched on from the Arctic tundra. So it seems that being a lover of meaty and fatty is not so bad—just as long as the animal sources for these foods have dined on the right things.

## Indigenous Diet Key Component #6

Naturally raised meat and dairy as a precious commodity:

- Meat in small quantities
- Organ meats and whey used in cooking
- Liberal use of proteins from nonmeat sources such as nuts, legumes, and grains

## Nonmeaty Fats and Oils

In a natural setting, our ancestors' innate attraction to fat would also have led them to choose healthy fats from nonmeat, unprocessed food sources such as nuts, seeds, grains, and fatty fruits (olives, palm fruits, avocados, etc.). These foods, when eaten whole or in a minimally processed form such as olive oil, offer us an excellent balance of fats (including omega-3, omega-6, monounsaturated fats, and certain beneficial saturated fats) as well as vitamins, minerals, and fiber. (See Appendix C for a more detailed discussion about these fats.) Processed plant oils such as corn, sesame, soybean, sunflower and canola oil tend to have a less optimal fatty make-up. These are used sparingly in most indigenous recipes. There is one basic reason for this: oil from these plants can only be extracted in large quantities by using big modern machines. Although archaeologists have discovered oil presses that date back over 5,000 years, it is only in the post-industrial era that vegetable oils have become so affordable and abundant. As we will see in the chapters that follow, the increased consumption of these processed fats and oils has had negative health consequences.

The predominance of minimally processed fats in traditional diets had another nutritional benefit. Because cooks did not have the oil to deep-fry or cook at high heats, they had to leave their pots and pans at a simmer. The result was that their ingredients had a chance to gradually meld together, producing a final product with rich, complex flavors. In addition, this slow-cooking method preserves the nutrient content of vegetables and grains and prevents fats (which include oils) from becoming transformed into unhealthy compounds. So we can see that in a traditional setting, craving meaty and fatty foods led to excellent dietary choices thanks to the *kinds* of fats and oils that were used and the *way* that they were used.

## Indigenous Diet Key Component #7

Nonmeat fats from whole nuts, seeds, grains, and fatty fruits; minimally processed oils such as olive, palm fruit, or coconut oil

## Instinct #3: The Power of Post-Ingestion

*Post-ingestive* is the term Paul Rozin uses to describe the third and final instinctual behavior that helped our ancestors select healthy diets. After testing a certain food, we have the ability to link symptoms we experience, be they positive or negative, back to that food. Most often, this physical feedback is given to us by the places where food makes its first impression: our stomach and intestines. Suppose I sample a fruit that looks inviting but then develop nausea, vomiting, and diarrhea several hours later. Chances are that I will steer far away from that particular fruit in the future. This is certainly the instinct that led the Peruvian villagers in Las Palmeras in the Amazon to eat the cashew fruit but avoid the nut. Their ancestors must have sampled the nut and then, hours later, suffered from stomach cramps and nausea caused by the urushio in raw cashews. The first cashew samplers in that community had engaged a wonderful protective mechanism to avoid taking a lethal dose of this food. (By the way, my Peruvian friends were greatly amused to hear that North Americans view cashews as a luxury food and actually go so far as to cook them to make them edible.)

Positive food experiences, however, tend to encourage us to resample a food item. When our ancestors first experienced the mood-enhancing effects of caffeine from leaves of the *Camellia sinensis* tea shrub or the flesh of a cocoa pod, chances are that they went back for more. This positive postingestive feedback is probably what led many cultures to discover fermented foods made from dairy, vegetables, or grains (fermented foods contain probiotics, microscopic bacteria that help protect against allergies, improve digestion, decrease bloating, and boost the immune system). Icelandic *skyr* (yogurt), for example, was most likely invented after someone mistakenly drank fresh milk that had sat around too long. Much to their surprise, rather then playing havoc with their digestion, it turned out to have a soothing effect. Fermented and pickled foods from other cultures, such as Greek yogurt, Cameroonian *fou fou* and millet beer, Korean kimchee, Japanese tempeh and *natto*, or German sauerkraut, were probably invented in a similarly serendipitous manner.

## Indigenous Diet Key Component #8

Fermented and pickled foods

The practice of adding spices such as garlic, onions, chiles, turmeric, and ginger to foods is also, at least in part, a product of post-ingestive learning. As we discussed, most tangy and spicy taste sensations are not innately pleasing to humans (just watch the twisted grimace of a baby as she samples her first hot pepper). However, our ancestors quickly discovered that adding spices to foods was an effective way to prevent a late- night belly ache either because these seasonings aided digestion or delayed food spoilage. In addition, many spices are anti-inflammatory, with some having effects that are dramatic enough to offer immediate positive post-ingestive feedback. For example, when it comes to treating the joint pains from arthritis, turmeric gives pharmaceutical-grade anti-inflammatories a run for their money. These post-ingestive benefits help to explain why tangy or bitter spices gained acceptance over time and even came to be regarded as enjoyable.

## Indigenous Diet Key Component #9

Healing spices

Although our ancestors might not have known about the existence of vitamins and minerals, there is a lot of evidence to suggest that foods' post-ingestive powers also helped them to decide which *combinations of foods* gave them the best nutritional boost. For example, someone who is deficient in a vitamin called thiamin will most likely experience mental confusion and problems walking. If she then eats thiamin-rich foods such as whole grains, beans, and lean meat, she will feel better almost immediately. While not all nutrients give such a rapid effect, this feedback loop might offer us a clue as to why groups like the Aztecs chose to treat their corn with lime or why other cultures developed culinary practices or food combinations that had not-so-obvious health benefits.

## Anatomy of an Indigenous Diet: The Nine Key Components

1. Foods that are local, fresh, and in season
2. Food cultivation techniques and recipes passed down through the ages
3. Food traditions:
   - Communal eating
   - Eating for satiety rather than fullness
   - Observation of fasts and other food rituals
4. Sugar from whole foods such as honey, fruits, vegetables, and whole grains
5. Salt from natural unprocessed sources such as fish, sea greens, and vegetables
6. Naturally raised meat and dairy as a precious commodity:
   - Meat and dairy in small quantities to complement vegetables, whole grains, and legumes
   - Organ meats and whey used in cooking
   - Liberal use of proteins from nonmeat sources such as nuts, legumes, and whole grains
7. Nonmeat fats from whole nuts, seeds, grains, and fatty fruits; minimally processed oils such as olive, palm fruit, or coconut oil
8. Fermented and pickled foods
9. Healing spices

## Where Did We Go Wrong?

Paul Rozin's explanation helped clarify many things for me. It was not just dumb luck, after all, that led to the creation of the healthy cold spot diets featured in this book. Rather, each diet evolved when people hundreds to thousands of years ago applied their three basic food-seeking instincts—imitative, sensory, and post-ingestive—in a natural setting. Regardless of where the traditional communities were located around the globe, their food-seeking behaviors pretty much guaranteed that they made optimal food choices, given the resources at hand. They also guaranteed that their diets, no matter how unique, all shared a common list of *key components* or a basic anatomy.

So what went wrong? Why are so many of us eating so poorly in our modern lives? Why are diet-related chronic diseases, such as heart disease and diabetes, continuing to increase worldwide? Before we begin to make positive changes, it is important to understand where things went awry.

# 3. A Diet Lost in Translation

"Well, our indigenous diet is certainly not a very healthy one." My mother, who had just learned about my book project, was referring to the foods she had eaten throughout her childhood. These were home-cooked by her own mother, my Nana. Born in Chodorov, a small village in the Ukraine, Nana was barely twelve years old when she first set foot on U.S. soil. Accompanied only by her older sister, she brought few material possessions to remind her of the Old Country. Perhaps her most direct link with the life she left behind was a mental inventory of recipes learned in her mother's kitchen in Chodorov. These same recipes, adapted to mid-century "modern" American foods, served as the backbone to Nana's cooking throughout her life.

When I picture Nana, I see her armed with a large mixing spoon and an apron in the open kitchen of her house on Cape Cod. Her guests are all seated at a long formica-topped dinner table as she hovers over them, piling huge servings of home-cooked food onto their plates. Occasionally, she gives an encouraging nudge with her spoon and chides, "You eat like a bird. Eat a little something." For all of us, Nana reigned supreme in the kitchen

39

as she served course after course of food with strange guttural-sounding names like borscht, kasha *varnishkes*, and *lochshen* kugel.

If these foods, as my Nana prepared them, were truly the Chodorov indigenous diet, then my mother was right. It was not a particularly healthy one—the proof being that many of Nana's contemporaries who ate these same foods (including my grandfather) suffered untimely deaths from heart disease, stroke, or diabetes. But was Nana's food really the diet of my ancestors in the Old Country, the meals that my great-grandmother Esther cooked over a wood-fueled stove and served up by candlelight in her home in Chodorov? As I learned more about the indigenous foods of Eastern Europe, I discovered that the dishes prepared in my grandmother's village in the early 1900s bore only a limited resemblance to those served in her New England kitchen sixty years later. While great-grandmother Esther's food provided an excellent array of nutrients and a balance of proteins, healthy fats, and unrefined carbohydrates, Nana's food was a real "heart-stopper." Something had happened as these recipes made their long voyage from the Old World to the New.

Take the borscht, for example. Both in Nana's kitchen and in Chodorov, "borscht" stood for beet soup. But this is where the similarity ended. In Chodorov, borscht was made of fresh grated beets that were simmered with water or a light chicken stock, flavored with salt and pepper, and adorned with chives and a dollop of yogurt or clotted cream. This borscht was ruby red in color. Nana's, by contrast, was a Pepto-Bismol pink concoction made with canned sweetened beets, generous amounts of sour cream, and Swanson's chicken broth. It also harbored a secret ingredient: a packet of Lipton's instant onion soup mix. From here Nana's meals continued to diverge from her ancestral ones: Wonder Bread instead of a whole-grain barley loaf, iceberg lettuce with mayonnaise and ketchup dressing instead of Brussels sprouts or cabbage, white pasta mixed with bulgur instead of straight bulgur, and so on and so forth.

This story of a transformed diet is certainly not unique. Most patients and friends I have spoken with feel that some time in the not-too-distant past their indigenous diet was mutated or lost altogether. In my Nana's case, the end result was a much less healthy diet than her indigenous one. However, this change for the worse seems to have happened almost across the board as evidenced by the fact that the rates of obesity, diabetes, and heart disease continue to climb in the United States and in most other parts of the industrialized world. Why is this the case?

Once again, the answer to this question lies in the three basic food-seeking instincts described in the previous chapter: imitative, sensory (with a preference for sweet, salty, meaty, and fatty), and post-ingestive. But here is the twist: while these food-selection impulses almost uniformly worked in our favor as we negotiated a pre-industrial world, they seem to have turned against us as we have transitioned into a modern way of life. In places as varied as Chodorov, Crete, or the remote mesas of Copper Canyon, these biological drives helped local people to maintain a diet with all the key healing components. Ironically, these very same instincts are leading most of us astray. Let us find out why.

## When Imitation Gets You in Trouble...

Take the instinct to imitate, for example. Imitation in a pre-modern setting ensures that each generation will build on the knowledge and eating traditions of the previous one. The end result is the healthiest (or least toxic) diet possible given the local food resources. By contrast, much to our detriment, most of us these days find ourselves imitating images in the media. In a world where large corporations compete viciously to get us to eat their manufactured foods, advertisers have learned how to tap into our drive to imitate. Everywhere we look, be it print magazines, billboards, television, the Internet, or movies, we see messages suggesting that attractive role models eating the advertised products. If you have any doubt about this, just check out the cereal ads aired during Saturday morning cartoons or the beer spots that accompany *Monday Night Football*. In each case, they show a group of beautiful people delighting in (and often fighting over) the food or beverage in question. Recently, a friend confessed to me that she has had to resist the temptation to pull into a Taco Bell and order a Taco Supreme. Her obsession with this fast-food establishment started after reading a *Vogue* interview with Angelina Jolie in which the actress announced that she liked to eat at Taco Bell.

At the same time that the mass media are pushing us to imitate celebrities, we have lost our opportunity to imitate our elders. It is now customary for people to leave home at a young age and live apart from their parents or grandparents, often being separated from one another by hundreds of miles. With this distance, we have lost our opportunity to learn how to grow, collect, and prepare our indigenous foods. In addition, the family meal during which a group sits down together on a daily basis is slowly becoming extinct. Most of us find ourselves eating alone and often in front of the television. This is a practice that invariably leads to overeating, since television prevents us from noticing our body's natural cues that tell us we are full. In addition, eating in isolation offers people little chance to learn table manners, share food lore, or appreciate the recipes of the previous generation. For many, celebrations involving feasting or fasting have also lost their importance. As a result, modest everyday meals punctuated by monthly or bimonthly feasts have given way to daily feasts where every meal becomes an overindulgence in calorie-rich foods.

## Sweet and Salty Out of Whack

Our love of sweet and salty once led us to select foods that had an optimal array of nutrients and fiber. Unfortunately, these sensory preferences are now causing many of us to develop chronic diseases such as diabetes and heart disease. Modern processed foods have a much higher concentration of saltiness or sweetness than foods found in nature so they really hit our natural response button. Furthermore, in the same size bite, processed foods are likely to give you many more calories and a lot fewer nutrients than their natural counterparts. Perhaps the most extreme example is soda pop, which offers a hefty dose of calories but, at the same time, does little to make us feel full. In fact, in one study, subjects who drank cola with their lunch ate just as much food as subjects who were served water or diet cola. In effect, the soda drinkers ended up consuming more total calories than the water drinkers because the soda drinkers did not reduce their food intake to make up for the extra liquid calories they had consumed.

While most of us can accept that soda, candy bars, and chips are modern processed foods with disproportionate amounts of sugar and salt, there are other less obvious foods that qualify for this label. Take a refined grain such as the flour in white bread. The process of refining grains entails removing the two most nutritious parts of the grain: the coarse husk and the oily inner germ. These two parts of the grain give baked goods a rough texture and nutty flavor, and their removal leaves only the white endosperm, a sweet fluffy form of starch with little nutritional value. (Even whole-wheat bread—as opposed to whole grain—is a highly refined food as it is made from flour that has retained some of the husk but none of the germ.)

This practice of refining grains to make white flour is by no means a recent development. In fact, archeologists have discovered grindstones for grain that date as far back as the Upper Paleolithic period. However, until recently, refined grain was considered a luxury item as the process to make it required that someone expend a lot of hard physical labor. First the grain needed to be stone-ground to remove the husk and then hand-sieved to remove the germ. All this changed in the later part of the nineteenth century, with the Industrial Revolution and the invention of the mechanized steel roller. Once food manufacturers figured out how to efficiently remove the oily but nutritious germ, they were left with a flour product that was much sweeter and had a guaranteed shelf life of months to years rather than days. Today most baked goods are made with this flour, offering us an array of inexpensive, calorie dense, sugar-packed, processed foods that can be enjoyed with little nutritional payback.

Along these same lines, many modern savory foods can harbor surprising amounts of salt when compared to high-salt traditional foods. If you sip a modest one-cup serving of your standard canned soup, for example, you have almost eaten your total daily recommended amount of sodium (2.5 grams). By contrast, a fairly large bowl of wakame seaweed will give you exactly half that amount of sodium. This is of particular concern since high-salt diets are linked to modern diseases such as hypertension, gout, and kidney stones.

## Meaty and Fatty to the Modern Extreme

The commercially available meat one buys in today's supermarkets is yet another form of manipulated food that plays havoc with our meat-seeking instincts. As we just learned from the Eskimos, our meat is only as good as what it has eaten. Most supermarket meat comes from animals that have been raised in pens and stuffed with large amounts of manufactured animal feeds, including processed corn, and soy. They are injected with antibiotics to prevent them from getting sick on these foods and they are fed hormones for the primary purpose of "beefing up" their muscle with fat. The end result is a cut of meat from an animal that is no more healthy than a human who spent a lifetime eating only junk food. Take a McDonald's burger, for example. Because of its excess fat content, it has a caloric density that is 90–100 percent greater than a similarly sized piece of wild game meat.

The sheer abundance of modern meat probably poses the greatest danger to us as meat-seeking creatures. In the United States, meat production receives generous government subsidies, and we have large farms that mass-produce our beef. Therefore, meat is readily available, and most of us place little value on a hunk of beef. Indeed, in most fast-food restaurants, ninety-nine cents will get you a burger, and many supermarkets offer daily specials on conventionally raised meat, selling it for as little as one dollar a pound. The closest most of us have come to hunting or foraging is to wander into the deli, butcher, or frozen food section of a supermarket and select a relatively affordable cut of meat to take home. What makes matters worse for us as a species of instinctual meat-eaters is the fact that the cheapest cuts of meat usually have the greatest concentration of chemicals and hormones and the lowest proportion of healthy omega-3 fats. As a result, the most readily available meat happens to be the least healthy product we can buy.

Alas, nonmeaty fats are yet another mine field. Vegetable oils such as corn oil, canola oil, palm kernel oil, peanut oil, and soybean oil are now the most abundant form of fat worldwide. In some respects, they also qualify as processed foods since many of these oils have been manipulated (as is the case of hydrolyzed or partially hydrogenated fats), extracted with acetone or high heat, or derived from a hybrid, new-fangled plant (such as canola). The end result is that these oils contain too much omega-6 fat and not enough omega-3 fat. (See appendix C for more information.) While any of these oils in moderation (with the exception of trans or partially hydrogenated fats) are certainly not a health hazard, they are inexpensive and produced in such abundance as to have an overwhelming presence in most diets. At this point, most epidemiologists and nutritionists agree that overconsumption of these oils has contributed to a dramatic worldwide increase in modern chronic diseases.

Even in Copper Canyon, one of the most remote communities that I visited for this book, I came across Mazola corn oil jugs lying empty and abandoned along mountain trails. Apparently the Tarahumara Indians were finding a way to lug these heavy containers to their isolated ranchos. While in the past, a fat-seeking instinct would have led them to eat small amounts of free-range pork lard, pick an avocado, or crave the naturally occurring sterols in corn, they were now able to satisfy this basic need with inexpensive vegetable oil bought at a local government store.

## Post-Ingestion Outwitted

The third and final instinct, which involves learning from our post-ingestive experiences, also seems to be getting us in trouble in a modern world. While our ancestors could usually depend on the fact that foods that made them feel good were good for them and foods that made them feel sick were not, this is no longer the case. Modern food manufacturers, recognizing that humans go back to foods that give them positive post-ingestive feedback, have turned this instinct to their advantage . . . and our disadvantage. Otherwise, why would soft drink companies put up to 72 milligrams of caffeine in a twelve-ounce soft drink? They have found that this additive improves drinkers' moods and, therefore, brings people back for more.

At the same time, in the name of preventing negative post-ingestive experiences caused by bacterial pathogens, these same manufacturers subject our foods to pasteurization, homogenization, and the addition of stabilizers and preservatives. These processes have added unnecessary (and sometimes harmful) chemicals to our food. At the same time, they have taken away much of what is healing and nutritive by killing protective bacteria and antibodies and denaturing vitamins and minerals.

One might ask, if our post-ingestive instinct is supposed to be a warning bell, why isn't it ringing loud and clear when we eat modern processed foods? The answer is that these new foods, by and large, have managed to outwit our alarm system. While I certainly have seen my share of patients who have allergies or sensitivities to food additives, most of us seem to be able to down a bag of chips or a soda without a recognizable negative reaction. As Paul Rozin explains, "In order for the post-ingestive system to work, it must detect negative effects such as nausea within hours of digestion. So slow poisons, foods that lead to atherosclerosis, etc., will go undetected." When you consider that the greatest health problems caused by these processed foods are chronic diseases that occur many years down the road, you can understand why our post-ingestive food selection system does us little good. These health problems tend to occur slowly and cumulatively over far too great a time span for this safeguard to be of any use.

## Where Do We Go from Here?

As I wrote this chapter, it amazed me to realize that the diet my relatives developed over generations in the Old Country was irrevocably altered when my grandmother crossed the sea and encountered modern foods. Of course, Nana herself is not to blame. The modifications that she made to her indigenous diet were done with the best of intentions. She was part of a whole generation that was led to believe that new and modern were truly better. It is only in retrospect that I see that her era in the kitchen dealt such a deadly blow to our traditions. Many friends and patients I have spoken to can identify a similar turning point in their food heritage.

Now we understand how we got here. What is next? Many of us have health concerns and want to eat in a more nutritious and healing way. Clearly indigenous diets in general have a lot to teach us. Do we each try to return to our own indigenous diet? Do we just eat from whichever diets happen to strike our fancy? Or are there certain indigenous diets that are better for us than others? In other words, with so many diets to choose from, should we be feeding our genes or our taste buds or focusing on a specific health problem?

# 4. Feeding Our Genes or Our Taste Buds?

When I first began to research this book, I assumed that my goal was to provide my readers and my patients with a roadmap to help them rediscover their ancestral diet. This seemed like a logical approach given that returning to one's dietary roots can have great health benefits. We see this not only in compelling stories like Angela's but also in larger scale nutrition studies. In places as varied as Mexico, Hawaii, eastern Finland, and the Australian Outback, indigenous people who were experiencing epidemic rates of modern chronic diseases have fared much better after returning to their indigenous diets.

Indeed, the idea that we have a specific set of foods that best matches our genes makes good intuitive sense. Traveling around the globe, I am struck by how varied we humans are as a species—at least in superficial ways. Most people in Cameroon look nothing like people in Okinawa, and both peoples look completely different from the majority of Icelanders. Many of these physical differences, such as skin color or hair growth patterns, have been explained by biologists as gene adaptations that help us thrive in our particular climate. Therefore, it stands to reason that the way that we react to food

47

would also be radically different depending on what foodstuffs have surrounded us for thousands of years.

It turns out this is at least *partially* true. Thanks to the human genome project, we have discovered that certain indigenous groups who have not experienced a lot of intermarriage or genetic mixing do have high frequencies of specific gene variations (also known as *polymorphisms*). These polymorphisms are thought to have occurred either by chance or because they offered some form of evolutionary advantage for that group. As one might expect, a certain percentage of these gene variations has something to do with how different groups process or metabolize food.

Perhaps one of the most widely discussed examples of a regional food-related polymorphism is *lactose tolerance*. A good proportion of the world's adult population experiences at least some digestive discomfort after eating dairy products. This is because they do not make lactase, the enzyme needed to digest lactose, a carbohydrate in dairy. However, people from northern Europe and some tribes in East Africa and the Middle East continue to be able to digest lactose into adulthood.

## A Theory About Lactose Tolerance

Certain food-gene adaptations are believed to have survived because they offered an evolutionary advantage. In the case of lactose tolerance, for example, some evolutionary biologists and geneticists believe that the trait is more common among groups whose ancestors were dairy farmers. If some of these ancestral farmers just so happened to have a mutation that allowed them to safely drink milk, they would have been at an evolutionary advantage. In other words, being able to digest the milk would have improved their nutrition and made them more likely than their lactose intolerant neighbor to raise children to adulthood. As a result of this advantage, over thousands of years, this lactose tolerance gene would become widespread in dairying populations.

Lactose tolerance is one of the best studied examples of food gene adaptations, but there are probably thousands of other food-related polymorphisms. Some of these are already known, and others have yet to be identified. Given this fact, it seems logical to conclude that the diet eaten by our ancestors will better match our genes and ultimately keep us healthier.

## Obstacles to Eating Your True Ancestral Diet

Theory is one thing, and reality is another. As I pursued this line of thinking, I kept coming up against some serious obstacles that would prevent most people from returning to their indigenous diet. First of all, there is the daunting task of figuring out who one's ancestors really are. While some of the patients I introduce in the following chapters identify themselves by one specific ethnicity, in reality their parents or grandparents come from many different parts of the globe. When I discussed this issue with Robert Nussbaum, MD, head of the division of genetics at the University of California, San Francisco, he confirmed that most of us in the United States are essentially mutts. Even among individuals who consider themselves to be in the *same* ethnic group, studies have shown that there is a wide variation in ancestry. As a result, it is nearly impossible for most of us to reconnect to one specific indigenous diet.

Problems of retracing ancestors aside, there are other obstacles to rediscovering one's own indigenous diet. Let's say you do happen to know the general region that your forefathers and mothers came from; does this necessarily connect you with a specific indigenous diet? My ancestors, for example, seem to have spent the past several thousand years living somewhere around the Black Sea. But I know that Soviet Georgians (my father's side) ate quite differently from people like my Nana who lived in the Ukraine. Furthermore, there are instances where indigenous groups living in the same geographic area eat different foods depending on the social norms of their own community. These issues, combined with the fact that most of us are now several generations away from the last of our ancestors who ate indigenously, make it near impossible to accurately trace our roots to one specific diet.

Next, there is the problem of finding the right ingredients. As traditional diet expert Harriet Kuhnlein explains in chapter 2, indigenous diets rely on local resources. Let us say that your ancestors are originally from Okinawa in Japan but have relocated to South Dakota. Suddenly the closest seaweed source (a staple food in Okinawa) is over 1,000 miles away—this hardly qualifies as a "local" ingredient. While you might get lucky and find some dried seaweed in your local store, there will need to be a shift in your daily eating patterns so as to include staples that are grown in South Dakota.

Finally, even if we were to identify our ancestral foods and have ready access to the ingredients, could we possibly limit ourselves to one type of cuisine? True, in the places I visit in this book, most people still seem to be adhering to one set of cooking traditions. However, in the past two generations, most westerners have become accustomed to a wide variety of cuisines and flavors. We start to feel like life is getting boring if we do not have Mexican on Monday, Chinese on Tuesday, and Italian on Wednesday. My kids have been known to complain if they are served food from one part of the world twice in one day: "Oh no! Mexican again. I already had it for lunch!" I realized that suggesting to someone that they limit themselves to one type of food for the remainder of their days might be a bit unrealistic.

## Learning from Nutrigenomics

I consulted with Jose Ordovas, PhD, the director of the Nutrition and Genomics program at Tufts University, to understand how much a person's genetic makeup determines what she should be eating. Ordovas is one of the pioneers in the relatively new field of nutrigenomics—the study of how specific chemicals in food interact with our genes. He also has a terrific knack for making the complex science of genetics understandable. "Certainly," he explained, "genes do play a role in how we react to our food. But how genes affect the big picture of our health varies person to person, depending on the combination of gene variations (polymorphisms) that we have and also how they enhance or negate each other. Just think. We all know some people who can put everything in their mouth and not gain an ounce while others eat nothing and gain weight. Then there are some people who have a 20–40 percent change in body fat without a change in lipids [cholesterol] while others have a 2 percent change and their lipids skyrocket. In each of these instances, that's their genes at work."

But when I asked him to give me a sense of how much of our total health is programmed in our genes as opposed to how much is determined by diet, exercise, etc., his answer surprised me: "On average, our genes contribute no more than 25 percent toward our total health picture." In summary, Jose Ordovas believes that for the majority of people, *our specific genetic makeup plays a relatively minor role in determining our destiny*, especially when compared to lifestyle factors such as diet and exercise.

Robert Nussbaum at the University of California put it more bluntly: "For most of the population, genes are just a small part of the health picture. This business of genes being the root cause of our health problems has resulted in a sense of powerlessness or lack of autonomy and is doing us all a great disservice." He went on to explain that even among groups that have a very high rate of a given chronic disease, genes seem to play a much smaller role than diet in determining the fate of each individual in that community.

When discussing the relative importance of genes and diet, both Nussbaum and Ordovas mentioned the Pima Indians of Arizona. The Pima have been extensively studied because they have some of the highest rates of type 2 diabetes in the world. Indeed, scientists have found that the Pima have a whole host of gene mutations that may relate to how they react to insulin, process sugars, and store fat. However, despite years of careful study and many theories, geneticists have not been able to pinpoint any one of these variations as being an obvious cause of the high rates of diabetes. In the end, most researchers seem to conclude that the true cause of these alarming statistics is a modern, industrialized diet that is high in sugar, refined vegetable oils, and processed carbohydrates. The case in point is that the Pima's genetic cousins, the Tarahumara Indians, who live several hundred miles south of Arizona in the Mexican states of Sonora and Chihuahua, have a low rate of diabetes. As we will see in chapter 4, the Tarahumara eat an indigenous diet and walk miles on mountain trails. Once again, it seems that lifestyle factors such as diet or exercise trump genetic polymorphisms.

So back to our initial question: will any old indigenous diet do? As I interviewed experts and listened to patients' stories, I realized that there is *no simple answer* to this question. On one side, it seems that any diet will do so long as it is not a modern one. Historically, rates of chronic disease really skyrocketed when people started eating processed foods. Yet, one cannot deny that genes do play a role in how our bodies metabolize our foods and whether these foods will increase (or decrease) our likelihood of developing a chronic disease. Perhaps even more important than the specific genes is how they manifest themselves in our family health history. Indeed, having a sibling or a parent who suffered a heart attack before age sixty probably increases one's own personal risk by 500 percent. Similarly, having two first-degree relatives with colon cancer before age forty-five may increase one's risk by 600 percent. I was in the midst of trying to sort out the relative importance of genes, family history, and personal health concerns when I had the good fortune to speak with Kauila Clark. Thanks to him, the issue of *who* should be eating *what* suddenly became clear.

It was a rainy winter morning in San Francisco when I called Kauila at his home on the island of Oahu. Despite the fact that it was 6:30 Hawaiian time, the energy in his voice suggested that he had already been up for a while, enjoying what was undoubtedly a lovely tropical winter day. *Kauila*, who is a respected elder in the community, otherwise referred to as a *kahuna*, straddles two worlds. In one, he is a traditional healer, and in the other, he is the chairman of the board of a medical clinic that serves the local community. In both the Western and traditional healing world, Kauila has one goal: to improve the nutrition of his fellow Hawaiians.

Kauila first became interested in nutrition decades ago when he noticed that the traditional Hawaiian foods that had nourished his ancestors were quickly vanishing from his native home. In the space of forty years, he watched Oahu transform from an island made up of little villages and taro fields to one covered with strip malls, exclusive beachfront subdivisions, and tourist hotels. Most of the undeveloped land is now owned by big agribusiness and has been planted with nonindigenous crops such as sugar cane. Canals dredged through these fields have led to coastal runoff of red dirt and other minerals. This, in turn, has killed off much of the coral and other sea life that surrounds the island. The octopi that used to be a staple food in Kauila's youth are nowhere to be found, and many of the traditional Hawaiian foods such as taro, breadfruit, fern shoots, purple yams, kukui nuts, fish, and free-range chickens are available only in expensive farmers' markets that are frequented mainly by tourists.

In parallel with these changes, Kauila has witnessed a dramatic deterioration in the health of his fellow Hawaiians. At this time, he estimates that over 35 percent of the adults in his community are morbidly obese, and a similar number suffer from type 2 diabetes and heart disease. "When you look at pictures of middle-aged Hawaiians 100 years ago," he laments, "they all had the bodies of twenty-year-olds. But that has all changed. . . ." Horrified by what he was seeing, Kauila dedicated himself to helping Hawaiians return to their traditional ways of eating. As a member of the board of directors in his clinic, he has supported agricultural programs that encourage local farmers to once again grow native foods. Similarly, he has collaborated with physicians like Terry Shintani to develop the Waianae Diet Project, a program which reintroduces people to their native diets.

However, it is probably in his role as a community leader, or kahuna (which translates literally to "keeper of the secrets"), that Kauila has made the greatest difference. As he sees it, there are some big obstacles to making dietary changes. Not only are the markets filled with modern convenience foods, but these also happen to be the only grocers that are subsidized by public assistance programs. Furthermore, Hawaiians, like all of us, are bombarded with messages on billboards and in the media that make processed, frozen, and fast foods seem desirable. Kauila told me that he realized early on that he could not compete with the advertising industry or the big chain grocery stores. "Instead," he explained, "I needed to find a story that would take people totally out of that realm so they could see that the issue of diet is something greater than themselves, something that also pays tribute to their ancestors. What I tell people is this: These [indigenous] foods are the foods that have ensured that you are here. You are a survivor in your family—your people made it for thousands of years on these foods, and you, right now, are the continuing link. What can you do to respect this process?"

In the past decade, Kauila has begun to see the fruits of his efforts. Traditional foods, once again, are becoming more available on the island, and there is a greater awareness among community members about the importance of eating indigenously. Those people who enrolled in the Waianae Diet Project have witnessed rapid reductions in blood pressure, blood sugar, and bad cholesterol. This has occurred without calorie counting—they have simply substituted nonindigenous foods with indigenous ones. As a result, the changes are pleasant enough that people are willing to stick with them over the long run.

But here is the catch: many of the people who have benefited from these changes are only part Native Hawaiian, and some are hardly genetically Hawaiian at all. Even Kauila, who attributes his excellent health to this diet, is not technically Native Hawaiian. In fact, he is part Irish, part English, part Portuguese, and, of course, part Hawaiian. But does one's degree of Native Hawaiiness really matter when it comes to determining who will do well on the Waianae diet program? According to Kauila, not in the least. What he has found is that if you enjoy the taste of Hawaiian food and have an emotional connection to the island, then you will do well on this diet.

## Diet Swapping

After speaking with Kauila, I realized that I had already come across a number of other promising examples of "diet swapping"—a term I use to describe when people adopt a diet that is very different from their ancestral one. One study looked at Anglo Celt Australians living in Melbourne. Anglo Celts who chose to switch to a Greek-style Mediterranean diet had lower rates of disease and lived longer than those who ate the typical Australian diet high in meat, refined starches, and dairy products. Interestingly enough, in this same study, the Anglo Celts who were eating a Mediterranean diet were compared to the Greek Australians who were eating a similar diet. Both groups seemed to do equally well, suggesting that this style of eating is transferable, offering health benefits regardless of whether or not your ancestors grew up around these foods.

Craig Willcox, PhD, an epidemiologist on the island of Okinawa and coauthor of *The Okinawa Program,* is also interested in the benefits of diet swapping. When I met him, he was launching a study to put U.S. military personnel on a traditional Okinawan diet. Although the study results have yet to be published, Willcox told me that he had seen a marked improvement in the participants' health as they abandoned the typical American foods that are served on the military base in favor of traditional Okinawan dishes. Finally, in a more extreme example of indigenous diet swapping, there are well-documented studies of Scandinavian explorers in the Arctic who have thrived for extended periods without vitamin deficiency by eating the traditional Inuit diet of fresh meat and whale blubber.

## Putting It All Together

Certainly genes and family history (and, of course, personal health issues) are factors to consider when you are deciding which indigenous foods to eat. Yet, as the diet swapping stories suggest, taste and connection to a place or tradition may be even more important. It is not uncommon for someone to research their supposed indigenous diet only to discover that there is very little about it that they find appealing. In fact, when I was testing the recipes in this book, one of my patients who describes her ethnicity as African American told me: "I really hope that I have some ancestors from Crete because those foods seem to sit with me a lot better than the West African stews." When I asked Robert Nussbaum whether a Mediterranean diet might work for this patient, he replied, "Maybe the foods this person has decided to eat happened to work with her genes. After all, she might identify as African American but have some Mediterranean ancestry." Then, after some thought, he gave an additional explanation. "Maybe [those foods] just made her happier. We all know that being happy, in itself, can improve our overall health."

Robert Nussbaum's and Kauila Clark's words helped me make peace with the idea that there is *no set formula* for helping you identify the indigenous diets that best meet your needs. As you read this book, you may use any number of factors to decide which foods to try first. You may be attracted to diets that match a part of your heritage. Of course, this will not work for everyone. This book only covers five regions, and therefore it is possible that your particular ancestry is not represented in these pages. If you are primarily concerned about a strong personal or family history of a certain disease you may want to first sample foods from the cold spot for that disease. It is important to note that while each regional diet is especially beneficial for preventing a specific health problem, it will also be helpful for maintaining overall good health. For example, the Tarahumara or Cameroonian diet will likely lower your risk of diabetes or colon cancer, respectively, but either diet will also help prevent heart disease and breast and prostate cancer. Finally, you may choose to do what I do. I browse through all the recipes and cook them on a rotating basis. Regardless of which approach you take, it is essential that you choose foods and recipes that match your taste buds and whose stories resonate in your heart. As we were saying our good-byes, Kauila left me with one last thought: "Those [fast-food] drive-throughs are way too appealing. . . . We need to find something very powerful to keep us away from them." Hopefully, he will agree that the next five chapters offer us just that.

Two

# 5. Copper Canyon, Mexico: A Cold Spot for Diabetes

I find it ironic that our first cold spot has the greatest physical proximity to the United States, and yet it was one of the most geographically inaccessible. Traveling little more than 400 miles south of the U.S. border, it took me two full days to reach Copper Canyon from San Francisco: first a plane trip to El Paso, Texas, then across the border to Mexico on foot, then to Juárez by taxi, and, finally, fourteen hours on increasingly rickety public buses that crawled up the Sierra Madre del Sur mountain range.

And why all this effort? Of course, I was excited by the prospect of hiking through dramatic rock formations and admiring firey sunsets. However, my true purpose was to learn about a diet that would help prevent a modern epidemic: type 2 diabetes. In the past seventy years, the prevalence of type 2 diabetes in the United States has increased over 700 percent, and the disease is slowly affecting younger and younger populations. While this is the case with people of all ethnicities, the most dramatic rise has been experienced by Latinos, Asian Americans, Native Americans, and African Americans. Furthermore, recent statistics have shown that diabetes is now taking center stage as one of the greatest health issues worldwide. In Mexico, for example, the diabetes problem has seemingly sprung up overnight with rates more than doubling in one decade. What is even more troubling is that, for every type 2 diabetic already diagnosed, there are two to three prediabetics who, if left untreated, will also develop diabetes.

Most people with diabetes or prediabetes probably have inherited specific genes that put them at increased risk for the disease. Nonetheless, lifestyle factors such as weight gain (especially an increase in belly paunch, or "visceral fat"), lack of exercise, and suboptimal food choices are the major factors that usually tip someone over into developing actual health problems.

In this chapter, my patient Arturo is trying to understand why his Mexican foods are making him prediabetic, while similar foods helped his parents and grandparents remain diabetes free. In search of an answer, I travel to a region near Arturo's ancestral village in northern Mexico. Along the way, I benefit from the research and wisdom of a diverse group of experts, including several preeminent nutrition researchers, a cook, and a gardener. Thanks to them, I discover traditional foods and recipes that, for thousands of years, have successfully helped prevent diabetes.

## More About Type 2 Diabetes and Prediabetes

Diabetes describes a number of diseases, all of which are brought about by problems regulating the hormone *insulin*. Insulin plays a vital role in controlling blood sugar levels. Most Americans who are diagnosed with diabetes have type 2 (previously called adult-onset) diabetes. It is hard to pinpoint one factor as the cause of type 2 diabetes; however, excess body fat appears to be a major contributor. This extra fat becomes a factory, producing abnormally high amounts of insulin and other hormones. Cells throughout the body quickly lose their sensitivity to these hormones, a condition referred to as "insulin resistance." Abnormally high insulin and blood-sugar levels then act as inflammatory toxins, damaging the kidneys, heart, arteries, and nerves. People with *prediabetes* have insulin and blood glucose levels that are higher than normal but are not yet high enough to meet the criteria for type 2 diabetes.

## Meet Arturo

"Twenty pounds and four ounces! How can it be? These days I feel like I eat less and weigh more." Arturo, a certified public accountant in his mid-fifties, was all about numbers. Today, he and I were sitting in my office looking at his chart and trying to figure out why his weight had shot up over twenty pounds in the previous six months. In addition, his fasting blood sugar, which had been skimming the upper limit of normal for several years, was now 114 mg/dl, a number that, by American Diabetes Association criteria, is solidly in the "prediabetes" zone.

Arturo was usually a light-hearted punster, but today he was unusually quiet. In fact, he looked downright morose. Trying to get at the cause of his weight gain, I asked him about exercise. While he did not feel that he exercised enough, he assured me that this had not changed over the previous year. Occasionally, he would make it to the gym, and, of course, there was always the weekend ride on his vintage Harley Davidson—an activity that he admitted did not burn many calories but was invariably the highlight of his week.

Finally, after a long silence, Arturo told me that he thought his real health problems started when he separated from his girlfriend Diana. Diana, who was part Venezuelan and part Mexican, was his companion on the weekend rides. She was also, to hear him describe it, a fabulous cook. She hated eating out and would concoct the most delicious meals using her Mexican grandmother's recipes. He fondly remembered her soups and stews. Sighing, he recalled her signature dish, pozole con pollo: the tender chicken, the rich tomato and ancho chile sauce, and of course, the pozole. To Arturo, there was nothing more delicious than those plump kernels of toasted corn simmered with chicken in a red sauce.

After the breakup, Arturo, whose mother's family was from Mexico, had maintained his love for Mexican foods. However, he was no longer eating at home. He had substituted Diana's meals with regular dinners at the local taqueria. He loved the spices and the beans and felt that, for a working bachelor who had no idea how to cook, the taqueria was a better choice than other fast foods. Looking for more clues as to how Arturo could have gained so much weight, I asked him what he routinely ordered. Occasionally, for a splurge, he would get the enchiladas or the chile relleno. However, his usual dinner was the super chicken burrito. Being conscious of his weight and blood sugar, he would always order the burrito without rice in order to make it "low-carb." "Plus," he added, "I usually get a small salad on the side."

Arturo was concerned about his weight gain and abnormal lab results and raised the issue of prevention. A self-described "pill popper," he asked if there was a medication he should use to treat his prediabetes. I explained that the research consistently shows that it is less effective to treat prediabetes with medication than it is to treat it with diet and exercise. Most prediabetics have fasting blood sugars that are slowly climbing to diabetic range over a period of five to fifteen years. During this time period, making lifestyle changes can prevent progression to diabetes and protect against damage to kidneys, arteries, and nerves. Similarly, these changes can also help early diabetics slow the progression to becoming even more severe diabetics. In one large study, gradual weight loss (with a goal of losing 7–15 percent of total body weight) and routinely getting at least thirty minutes of vigorous exercise five days per week returned blood sugars to a normal range and reduced the risk of developing diabetes by almost 60 percent.

## Confusing Advice

While Arturo understood that weight loss needed to be one of his ultimate goals, he said he was terribly confused about the best antidiabetic diet. Low carb? Low protein? Low fat? Not too long ago, it seemed that most packaged foods and diets proudly bore a "low-carb" label. Even McDonald's was promoting a low-carb meal. Recently, Arturo had been advised by a nurse friend that it might be best if he cut carbohydrates from his diet altogether since carbohydrates are broken down into sugar—one of the main culprits in diabetes. Despite the popularity of the low-carb approach, Arturo was not so sure that this made sense. After all, his mother, aunt, and grandparents ate a very high carbohydrate diet consisting of rice, beans, and tortillas for most meals and lived to ripe old ages without diabetes. Plus, he had already tried an Atkins-type high-protein diet with minimal success. He told me that the food plans worked well for about a month, and then he began feeling uncontrollable cravings for tortillas and rice. After six weeks, he caved in and started to binge on these foods. Certainly this binging did not help his weight and blood sugar situation.

More recently Arturo had seen a variety of food products on the supermarket shelves that bore the label "low-glycemic index" with a note that they are good for diabetics. This intrigued him, and he asked me what this meant and whether he should be eating these foods. I explained that the glycemic index (GI) is not such a new concept. In fact, it has been widely discussed in nutrition circles since the early 1990s but has only recently made its way into popular nutrition, joining "low-carb" and other packaging buzzwords. The glycemic index is a shift away from the "carb is a carb is a carb" concept to the idea that different types of carbohydrates can have different effects on blood sugar. The less a given food raises blood sugar, the lower its glycemic index. For example, jasmine white rice (GI=109) has almost four times the glycemic index of another staple grain, pearled barley (GI=29). Research has shown that total calorie consumption being equal, low-glycemic diets seem to control blood sugar better than high-glycemic diets.

While the glycemic index of specific foods is certainly a useful concept, it only gives us one part of the picture. The way foods are prepared, how ripe they are, and what they are mixed with also determines the speed with which they release sugar into the bloodstream. For example, a serving of pasta cooked al dente has a lower glycemic index than a serving of mushy, overcooked pasta, and white bread alone has almost twice the GI of a popular Scandinavian snack, white bread dipped in vinegar. Furthermore, glycemic indices give you no idea about the overall nutritional value of a food. For example, a can of Coca-Cola has a glycemic index of 63, while five dried dates have a glycemic index of 103. You guess which has more vitamins, fiber, and other nutrients. Finally, because the glycemic index is related to the amount of digestible carbohydrates in a food, it can be misleading. Foods that are mostly water, fiber, or air (such as carrots) will *not* cause a large increase in blood sugar levels even if their glycemic index is quite high.

## Indigenous Foods: An Rx for Preventing Diabetes

Rather than giving Arturo a list of foods or a table of glycemic indices, I began to search for a specific diet that had proven effective for preventing diabetes and prediabetes. Amidst the dozens of articles about low-carbohydrate, low-fat, and low-glycemic diets, there was one older journal article that really stood out from the rest. It was titled, "Slowly digested and absorbed carbohydrate in traditional bushfoods: a protective factor against diabetes?" The authors (who happened to include Jennie Brand-Miller, PhD, the author of the national best seller, *The New Glucose Revolution*) were struck by how rapidly Pacific Islanders and Australian Aborigines were developing prediabetes or full-blown diabetes once they switched from their native high-carbohydrate diet to a Western high-carbohydrate diet. The authors wondered why the starchy carbohydrate native foods that the Pacific Islanders and Aborigines had eaten for generations had a protective effect against diabetes, while the starchy carbohydrate Western staples seemed to bring on the disease. The native foods included a variety of nuts, roots, and seeds with unusual names like cheeky yam, black bean seed, and bush onion, while the Western foods included russet potatoes, white bread, white pasta, mass-produced corn, and white rice.

### Some Indigenous Slow-Release Carbohydrates from Around the World:

Stone-ground flour; cracked wheat, rye, or barley kernels; quinoa; whole-cooked legumes; steel-cut oats; and millet pearls. See Appendix B for more details.

To solve this mystery, Brand-Miller and her colleagues put the indigenous and Western foods through a rather odd testing process. They collected human saliva and mixed it with standard quantities of each food. Then they measured how quickly the starches were broken down or digested by this saliva cocktail. It turns out that the indigenous carbohydrate foods, whether they were grains, legumes, or tubers, had one thing in common that differentiated them from the Western carbohydrates: they were all harder to break apart and digest. As a result, these traditional foods produced a smaller (and less rapid) change in blood sugar and insulin levels than the Western ones. The researchers concluded that the indigenous foods were slow-release foods and that these kinds of traditional staples could be recommended as part of the dietary treatment of diabetes.

Finally, I had found a series of articles about nutrition and diabetes that made sense to me! Forget counting carbohydrates or measuring glycemic indices—I had a hunch that if I could understand what Arturo was eating, I could recommend some substitute slow-release traditional Mexican foods that appealed to his palate.

## Five Reasons Why Slow-Release Indigenous Foods Are Antidiabetic

- Slow-release foods are slowly digested. This prevents an overproduction of insulin and keeps blood sugar levels from going too high or too low.
- Slow-release foods are fiber-rich and give you a satisfied full feeling, cutting down on hunger between meals. This way you don't crave the fast-release snacks that can raise your blood sugar.
- Slow-release foods are nutrient-rich. Indigenous foods are unrefined, or whole grain, and therefore they retain their vitamin and mineral content. These nutrients are often lost when grains are refined. For example, when wheat is refined

to white flour, only 15 percent of the magnesium is preserved. Low blood magnesium levels are linked to insulin resistance, poor blood-sugar control, and diabetic complications.

- Slow-release foods are free of added processed fats and oils (including partially hydrogenated or trans fats) and contain healthy plant-derived fats called stanols or sterols. Stanols and sterols *lower* unhealthy blood triglyceride levels, while added fats often *raise* these levels, causing more body fat buildup, inflammation, and poorer blood-sugar control.

- Slow-release foods have unique antidiabetic properties. There are some traditional foods that actually seem to make cells more sensitive to insulin. These include some herbs and nopales (prickly pear cactus pads). See the nopales and herb paragraphs of this chapter for more details on insulin-lowering foods.

## Arturo's Burrito

I know Arturo's favorite taqueria well and have stopped in for a meal from time to time. But one particular day, I went in with the specific goal of doing some important research. I ordered the super chicken burrito with no rice and a salad. Like a bench scientist, I delicately began to disassemble it trying to understand its constituent parts. Under the tightly wrapped tin foil cylinder, I found a large flour tortilla that unfolded to the size of a medium pizza. Inside were huge glops of sour cream and guacamole. Moving these aside, I uncovered a trench of glistening refried beans, which the cook behind the counter informed me were Peruana beans (an off-yellow dried bean) that had been boiled for several hours then left to simmer in a generous helping of vegetable oil and salt. Lying in the trench was a long line of melted cheddar cheese and shredded chicken. Sprinklings of red salsa with onions intermingled with the other ingredients. The salad was much easier to study: a small pile of iceberg lettuce with two slices of tomatoes and a dollop of a creamy white dressing.

I rerolled the burrito and took a bite. The taste was salty, peppery, and meaty, the texture creamy and chewy, and there was a touch of sweetness and crunch, thanks to the salsa and onion. I had intended to eat half, but all of a sudden discovered that I had scarfed down the whole super thing with several handfuls of tortilla chips to boot. I had done my homework; I now understood why Arturo had gotten himself into trouble.

## In Search of a Native Diet for Arturo

Arturo considered this diet of burrito with the occasional enchilada or chile relleno to be his native food. But was it? After all, he was eating in a Mexican establishment, and most of his fellow diners were also first- or second-generation Mexican Americans. He was puzzled by why this food had made him so unhealthy. Once again, he gave the example of his mother, born and raised in a tiny village in the Mexican state of Chihuahua, who had eaten Mexican food her whole life and was doing just fine. We talked about this at length, and both agreed that we needed to discover what, if anything, she had done differently. We wanted to learn about foods that could help get Arturo off the downhill slide to diabetes and on the path to wellness.

Sadly, modern day Chihuahua, which borders Texas and New Mexico, has very little to teach us about diabetes prevention. Here soft drinks, potato chips, packaged instant soups, and pastries have rapidly replaced traditional foods, turning Chihuahua (along with many other parts of Mexico) into one of the world's great hot spots for diabetes. Interestingly, it is those parts of Mexico that border the United States that have the most alarming statistics, almost as if diabetes were a contagious disease that was slowly spreading from country to country.

Within this ever-expanding diabetes zone, there happens to be one notable cold spot. It lies 100 miles west of Arturo's ancestral village, in a vast series of deep canyons known collectively as Copper Canyon, or *Barrancas del Cobre*, because of its rich mineral resources. This is home to more than 50,000 Tarahumara Indians, many of whom have continued to thumb their noses at the modernization and industrialization experienced by other parts of Mexico. John Kennedy, an anthropologist who spent several years living in the Copper Canyon area, tells the remarkable story of the Tarahumaras' resistance to assimilation in the face of constant onslaughts from missionaries, mining companies, loggers, and commercial enterprises. More recently, their remote communities have been assailed by an even larger threat: the tourist trade. To preserve their traditional way of life, many Tarahumara clans have retreated high into the craggy Sierra Madre del Sur mountain range. They use what tiny bits of flat and fertile land they can find to build their mud homes, grow their crops, and tend their livestock. By sheer perseverance, many of the clans have maintained their traditional culture, including the social structure, ceremonies, language, farming techniques and, perhaps most importantly, indigenous diet.

One of the first people I contacted to learn more about the Tarahumara was William Connor, MD, a professor of nutrition at Oregon Health Sciences University in Portland. Starting in the 1960s, Connor became intrigued by the stories he had heard about the Tarahumara Indians of Chihuahua, Mexico. The fact that they were a peaceful, egalitarian society that valued living in harmony with nature took his fancy. However, what really fascinated him was that the Tarahumara (or Rarámuri as they call themselves) represented a cold spot for diabetes and heart disease.

For several summers in a row, William Connor, accompanied by a small group of other researchers, traveled to Copper Canyon to learn about the diet and lifestyle of the Tarahumara. During these visits he also collected blood samples to see if the rumors were indeed correct regarding the scarcity of diabetes and heart disease in their communities. So what did Connor and his colleagues discover during their time in the Canyons? I called him at his lab in Oregon to get a first-hand account of his Tarahumara nutrition studies.

In actuality, the Tarahumara were not as healthy and hale as William Connor had expected, as they suffered from many diseases commonly caused by poverty, lack of medical care, and rough living conditions. These included infections, injuries, and lung problems (resulting from years of cooking over open fires). But what about the three common killers in more modern cultures: diabetes, heart disease, and cancer? Extremely rare. In fact, Connor saw no evidence of these health problems when he performed physical exams on community members, and the blood samples that he had collected confirmed these impressions.

Many of us may be skeptical when we hear this. Are the Tarahumara simply dying too young to develop modern health problems? Connor had asked the same question and was rather surprised at what he discovered: You see, typically in the United States and other industrialized countries, blood sugar, blood pressure, and cholesterol levels slowly creep up as people age. Among the Tarahumara, however, there was no such creep in numbers. The seventy-plus-year-olds that Connor studied (and there seemed to be plenty of them around) had the same LDL cholesterol levels and the same blood sugar numbers as the twenty-five-year-olds!

One could write these impressive laboratory findings off to lucky genes, meaning that the Tarahumara were less susceptible to these modern diseases. However, as we discussed in the previous chapter, the Tarahumara are distant cousins to the Pima Indians of Arizona, a group that has more than their share of type 2 diabetes cases. Genes seemed to have very little to do with the lack of diabetes amongst the Tarahumara. I was now convinced that it was worth my while to take a trip to Copper Canyon and see what they were eating.

## Those Who Walk Well

Like many other traditional societies, the Tarahumara get lots of points for exercise, and of course, this must play an important role in their good health. *Rarámuri* literally means "those who walk well" in the Tarahumara traditional language. Long-distance running or fast walking through the mountains to tend farms, herd livestock, or visit friends has become a trademark of the Tarahumara culture. In addition, there are still occasional *rarajipari* competitions. These are essentially long-distance kickball races where runners cover enormous distances barefoot while kicking a baseball-sized wooden ball. The object is for the runner to control the ball until he reaches the finish line, which may be a hundred miles away!

Getting off the bus in Creel, a medium-sized mountain town that serves as the gateway to Copper Canyon, my heart sank. Plastered on every spare surface were larger-than-life advertisements for Sprite, Coca-Cola, and Tecate beer. My family and I had traveled over two days by air, foot, taxi, and bus just so that I could learn about the native diet of the Tarahumaras. And here we were, greeted by soft drink and beer ads at the end of the line! To make matters worse, minutes after our arrival in town, a one-legged vendor rumbled up in a wheelchair to show me his wares, the basket on his lap filled with an assortment of packaged candy, cigarettes, and chips. After some small talk, he mentioned that he had lost his leg to diabetes. Certainly, this was not the cold spot described by Connor in his research.

It was then that I recalled more details of my conversation with Dr. Connor. He had mentioned to me that Creel, located at the top of the canyons, was not the place to experience a real slice of Tarahumara life. Crawling with backpackers and *mestizos* (non-Tarahumara Mexicans) and home to the more assimilated Tarahumaras, this was just another bustling frontier town. In fact, despite its remote location, it was only a matter of time before Creel would have its first strip malls and fast-food establishments. If I wanted to get a glimpse into authentic Tarahumara culture, then I needed to get out of town and head to the remote ranchos that lie deep within the canyons.

So down we headed, inching our rented pickup along an endless series of unpaved switchbacks cut into the canyon's sheer wall. Our destination was the tiny hamlet of La Bufa, which lies three-quarters of the way down the steep descent into Batopilas Canyon. There Sherry and David, a couple from Texas, had rented us a room in their small hillside house. Our plan was to use this as a base camp for hiking and exploring the area and for visiting Tarahumara ranchos.

At first glance this might seem like a reasonable plan; however, I quickly learned that hiking up or down on the footpaths that lead to the ranchos was a death-defying act. I usually consider myself to be a fairly capable hiker, but these pencil-thin, uneven trails left me in a constant state of terror. With any misstep or stumble, one risked falling to a certain death on the canyon floor, thousands of feet below. At times, after hours of picking our way along an ill-defined trail, my family would finally come across a rancho set high on a mesa or tucked into a rocky overhang. From a distance, we could catch glimpses of daily life: a blackened pot cooking on an open hearth; a woman picking nopal leaves; goats, chickens, and donkeys grazing; and children playing. Always the Tarahumaras would wave, and their dogs would offer us a curious sniff, but never would they beckon. So we did not venture any nearer. My only face-to-face contact with a Tarahumara was as he or she nimbly passed me on the trails, me edging along, knees chattering, and hands clinging to the manzanita shrubs with all my might—hardly an opportune time to discuss food and recipes. Obviously getting off the beaten track and learning about indigenous foods of the Tarahumaras was easier said than done. The anthropologist John Kennedy spent months hanging around before anyone even invited him into a rancho or to a *tesgüinada* (a traditional home-brewed beer party). Given the multitude of threats that they have endured, their stand-offishness was quite understandable. Even so, I was left wondering what I could possibly hope to accomplish in my ten-day junket.

## Maria and Her "Three Sisters"

I was about to give up on my recipe-finding expedition when I got lucky and met Señora Maria Cruz. After enjoying a delicious meal of fresh pan-fried mountain trout, peppers, black beans, and handmade tortillas at a rustic lodge near Cusarare, I decided to wander into the kitchen and ask for some recipes. Inside were two women, both dressed in the traditional orange and purple embroidered skirts, their hair hanging down their backs in thick glistening braids. Neither one spoke very much Spanish, and my attempts at striking up a conversation made them giggle and whisper to each other. I pulled out my notebook, hoping to jot down some ingredients, but they moved away eyeing me suspiciously. Finally, in a last ditch effort, I picked up a ball of *nixtamal* (fresh ground masa corn) and tried to form it into the shape of a tortilla. With this, one of the women came over, seized the dough, and proceeded to show me the proper way to form a tortilla.

Apparently the international language of cooking was a good icebreaker. Soon we were chatting away using some Spanish and lots of sign language. My tortilla instructor, Maria Cruz, was thirty-two years old and had grown up in a traditional rancho not far from Cusarare. Her husband had died several years ago, and she decided to take a job so that her son could have an opportunity to attend school. Even though she now worked for a tourist establishment, she still preserved many of her Tarahumara customs, including her childhood recipes. Maria Cruz said she was raised on the *tres hermanas*, or "three sisters,"—corn, beans, and squash—supplemented by eggs, chicken, chiles and gathered herbs, nuts, berries, wild greens, cactus, seeds, and a variety of fruits, including oranges, tomatoes, and avocados. Occasionally she would have wild game or fish. As she spoke, I looked out the window at the high sierra landscape, with scrubby pines and rocky earth, and I was amazed that all the foods that she described could be grown in such a harsh environment.

I asked Maria about the goats, pigs, and cows that I saw grazing around the ranchos. She "tsked," shaking her head. These animals are slaughtered only on special occasions, such as weddings, harvest celebrations, communal work days, or *Semana Santa* (Easter week). After all, livestock are a savings account and a protection against starvation; once you eat them, their value is lost. Better to use the animals to fertilize the fields and improve their yield of corn, squash, and beans, and better to eat the eggs rather than the chickens.

The next morning Maria served us eggs, salsa, and fresh tortillas, proudly announcing that the salsa was her mother's recipe. The combination was delicious, and I decided that only a complicated, multistep recipe could produce a breakfast that was this tasty. In fact, when Maria gave me the directions and the list of ingredients, it was surprisingly simple. The excellent flavor was all in the scratch-fed, free-range eggs, the small jalapeño peppers, and the fresh cilantro. Many of the stew recipes that she shared with me (one-pot cooking is definitely the mode for most traditional Tarahumara foods) were equally simple, but the flavor of the dishes was a testament to the importance of good ingredients.

Luck struck twice, and several days after meeting Maria Cruz, I was introduced to Señor Taurino. Taurino held a number of odd jobs in and around the canyon and also worked part-time as Sherry and David's watchman. When I first met him, it was in the middle of *Semana Santa*. He was sitting outside his one-room cabin and playing a goatskin drum that he had made especially for Holy Week. From high in the cliffs, I could hear someone else echoing his rhythm in a call-and-response pattern. It was a blistering hot day, and I was surprised to see that he was sipping a fiery orange brew that was still steaming. He called it *alí* and told me it was a celebration drink made from cilantro, garlic, lime, red peppers, and the scale left behind by some kind of tree insect. He offered me a sip. Once I got over the idea that I was drinking insect droppings, I found it to be a sweet and spicy delicacy.

Traditionally, it is the Tarahumara women who do most of the cooking, but Taurino was an exception. Having been a bachelor for years, he had developed a good deal of kitchen know-how and considered himself to be an accomplished chef and gardener. As my kids took turns trying out his drum, Taurino and I sat and talked about his garden, his outdoor kitchen, and his favorite recipes. Taurino gardened in a small 20 × 30-foot space but seemed to grow everything that he needed, including several stalks of maize. There was a dense bed of cilantro, which he regarded more as a green than a mere spice. Surrounding the cilantro were clumps of scallions, tiny red peppers, and garlic. A variety of fruit trees were scattered throughout the patch, laden with papayas, grapefruits, avocados, and pomegranates. Large gourds were growing helter-skelter on vines (both edible and for making bowls) and bean stalks were trained against a fence.

For Señor Taurino, the trick to making delicious food is to use fresh spices and to really let your beans soak up the flavor. "Beans are only as good as their neighbors," he cautioned. His outdoor kitchen consisted of a gas burner, a stone metate (mortar) for grinding corn, and a rustic table. On the table was a row of glass jars, each filled with a different dried spice. Taurino only had one knife, one wooden spoon, and one cast-iron cooking pot, but this seemed to be more than enough equipment for preparing a delicious Three-Sister Stew. Watching Taurino prepare his meal, I thought of how pleased Arturo would be to hear about the one-pot situation. Arturo had insisted that he would only cook a meal if it was simple to make and did not leave him with a pile of dirty dishes—this one certainly fit the bill.

## The Antidiabetic Garden

I asked Señor Taurino if he had ever met anyone in the canyon with diabetes, and he answered, "In town, yes; in the hills, hardly ever." Unprompted, he proceeded to point to all the foods in his garden that he knew prevented diabetes. These included the three sisters (corn, beans, and squash) as well as jicama. In the distance far outside the border of his tended garden, he pointed to a clump of nopal cactus. Indeed, scientific research has shown that cactus pads can lower blood sugar and help prevent diabetes. It was remarkable how this man, who had no formal education and had never worked in health care, could discuss nutrition with such authority. Listening to his comments, something suddenly dawned on me. No wonder the Pima and other indigenous Americans have been cursed with such high rates of diabetes! The foods that they have eaten for centuries are all slow-release, and some actually had special *anti*diabetic properties. It must be a terrible shock to their systems (and to their genes) to suddenly switch from these protective foods to a modern diet full of refined flour, sugar, and added fats.

### Some Tarahumara Slow-Release Foods

- Corn (maize)
- Beans
- Squash
- Jicama
- Nopales
- Herbs and spices

# Slow~Release Sister #1: Maize (Corn)

Maize is a food that is so central to Tarahumara cooking that, according to Taurino, many of his countrymen assign its seeds a divine power. In truth, eating in Copper Canyon is a gastronomical adventure in maize. After ten days there, I came to appreciate the subtle differences in tastes between the varieties and the enormous range of foods that can be prepared with this one starch. Depending on whether it is green or yellow, how much it is roasted, how much it is ground, or what it is mixed with, maize has a different flavor—and a different name. One day, as I headed out on a hike, Taurino offered me a reused Coca-Cola bottle filled with a light brown, thick liquid. This turned out to be *pinole*, a shake made from ground, toasted corn which is the Tarahumara equivalent of Gatorade and a PowerBar rolled into one. Pinole is traditionally carried in a goatskin as an energy booster for the long hikes through the canyons. It was surprisingly tasty and refreshing.

While maize is used in a variety of appetizing dishes, it is the Tarahumara tortilla that showcases this grain in its most delicious form. Typically, I do not think of corn tortillas as a slow-release food, but the tortillas that I tasted in Copper Canyon were a whole breed unto themselves—sweet, nutty, a little grainy, and oh so filling! The maize used to make these staples is the offspring of corn that has been growing in this same area for hundreds of years. If you look at the kernels, they are small and gnarly. Nothing like the plump, pale yellow, perfectly uniform corn kernels that so many of us are used to seeing in our supermarkets. It is easy to see that they are non-GMO (genetically modified) and packed with nutrition.

Not only were Taurino and Maria using indigenous ingredients, but they were also using traditional techniques to prepare the corn and make their tortillas—methods that were originally developed by their ancestors several thousand years ago. The corn is dried, boiled, and soaked overnight in a mineral lime solution to remove the outer shell. The inner kernel is then hand ground on the metate stone to create a dough (*nixtamal*) that is shaped into tortillas. Unlike the commercially made corn or flour tortillas that we can buy in most supermarkets, this ancient process creates a calcium- and niacin-rich food. Nutrients are preserved both by using more regional varieties of corn and by avoiding all the chemical preservatives, hydrogenated oils (trans fats), and fillers that one finds in most commercial tortillas. Because the maize is more crudely ground, I discovered that one handmade tortilla is almost as filling as two factory-made ones. In fact, when eaten with beans, studies show that the digestion rate of traditionally prepared corn tortillas actually slows to match that of the slow-release bean.

For most of us, getting a tortilla whose nutritional value approximates a Tarahumara tortilla can be a bit challenging. Living in San Francisco, I am lucky because I can walk into the La Palma Mexicatessen on the corner of Twenty-fourth and Florida Streets and join the line of loyal customers waiting to buy fresh ground maize (masa). If you live in an urban area with a high concentration of Latinos, you may have your own version of La Palma. Otherwise, you will have to settle for shopping in your nearby health-food store for dried masa, which you can make into tortillas using the recipe in this book. Alternately, you will often find that there is someone local who is producing fresh, handmade tortillas. If not, chain stores such as Whole Foods and Trader Joe's sell whole-grain corn or wheat tortillas that are a good alternative.

## How to Choose a Slow-Release Corn Tortilla

- Make sure it has at least 3 grams of fiber per tortilla.
- Make sure that it has been treated with lime as this enriches its nutrient content.
- Try, whenever possible, to select organic corn.
- Avoid partially hydrogenated fat or added vegetable oils.
- Avoid preservatives such as sodium benzoate. (Of course, without preservatives they will go bad faster, a problem you can solve by keeping them in the freezer and defrosting when you need them.)

## Slow-Release Sister #2: Beans

The beans whose vines grew helter-skelter around Taurino's garden patch are another important Tarahumara slow-release food. They help to control blood-sugar and insulin levels in a number of ways: First of all, they are low in calories and high in nutrients and fiber. Second, they are exceptionally filling, and a relatively small helping seems to go a long way. Because of these health properties, it is no surprise that legumes play a central role in most indigenous diets worldwide. There are over fifty species of beans that are native to the Americas. Many still grow wild as weeds. Inside the Batopilas and Urique canyons, I found scarlet runner bean growing wild, its long tendrils and gorgeous flowers leaping over bushes and up huge granite boulders. Beans (legumes) of any variety along with squash and corn are a terrific trio. I love them because they taste just right together, and I also appreciate that the combination gives an excellent balance of nutrients, fats, and proteins. Of course, as Taurino cautioned, beans are only as good as the spices that accompany them.

### Beans: Cook Your Own or Eat Them Canned?

My vote: Eat home-cooked dried beans over canned when you have a choice. Cooking dried beans is easier than you think (see Appendix B), and home-cooked beans offer a number of advantages.

- They tend to be fresher and more flavorful than canned beans.
- They are much cheaper, pound per pound, than canned beans. (Cooked dried beans are approximately ten cents per serving while canned are one to three dollars per serving.)
- They are more slowly digested than canned beans.
- They have less salt and fewer preservatives than canned beans.

If you simply can't cook your own—don't worry. Low-sodium canned beans are still a very healthy option, especially if you give them a good rinse. Both home-cooked and canned beans are excellent sources of nutrients and fiber.

## Slow-Release Sister #3: Squash

My kids found squash gourds in Taurino's garden that were so massive that it took two of them to lift them. Winter gourds (hard squash) and summer squash have been eaten in the Americas for several thousand years and are the third essential ingredient in the Three-Sister Stew. Although each squash variety has unique nutritional attributes, most squash are excellent sources of potassium, fiber, and beta-carotene (if orange or yellow in color). They also supply a fair amount of magnesium, B vitamins, and vitamin C. When you are selecting zucchini and other summer squash, you want to steer clear of the big mamas and go for ones that are smaller and more tender. Whenever someone tells me that they hate squash, it is usually because they have only tasted the kind that are overgrown and bitter. My preference is for yellow squash and green zucchini that are no bigger than a cucumber.

## Other Slow-Release and Antidiabetic Offerings from Taurino's Garden

Lying on the table near Taurino's metate was a collection of tan-colored hairy tubers. These were jicama, yet another example of a perfect slow-release food. Rich in the carbohydrate amylase, its sugar enters the bloodstream at a nice steady rate. My friend Dashka once told me that she carried a four-pound jicama eight miles into the wilderness to use as food for a backpacking trip. At the time, I thought it terribly odd to lug something so heavy into the woods. However, when I was offered jicama in the heat of the canyon's midday sun, Dashka's survival food of choice suddenly made perfect sense to me. As a filling and refreshing food, jicama is worth its weight in gold. Its hairy external peel is easily pared away, leaving the white, cool, juicy flesh. Jicama has such a subtly sweet flavor that hardly anyone finds it offensive, and it can easily switch hit between a savory or sweet dish. In addition, it can sit out at room temperature for weeks, hence it is an excellent camping food.

## A Perfect Slow-Release Snack

Taurino chops his jicama into french fry–size pieces, marinates them in lime, and then serves them topped with chili powder. An alternative is to marinate jicama in orange juice with a squeeze of lemon to mimic the bitter/sour taste from Mexican limes and oranges. I like to slice it into pieces and eat it raw with guacamole or salsa as a healthier (and more filling) alternative to tortilla chips.

The nopals that grow at the periphery of Taurino's garden are another excellent antidiabetic food. According to him, everyone in the canyon knows that these are especially good for controlling blood sugar and that it is best to eat the small, tender, bright green ones. Nopal, the paddle from the prickly cactus, grows everywhere in the deserts of Mexico and the Southwestern United States and is a mainstay in the Tarahumara diet. Why it is antidiabetic is still somewhat of a mystery. Some researchers think it lowers blood sugar by mimicking the hormone insulin. It is also a good source of soluble fiber, which may help slow the release of sugar into the bloodstream. Its juice has a slimy consistency, much like okra; however, if you grill it, the sliminess disappears.

The first time that I tried to pick and deprickle my own nopales, I ended up with a palm full of nasty cactus spines. It is hard to believe that such a delicious and nutritive plant could be so inhospitable. Since then I have learned some deprickling techniques and have included them in the recipe section for those who are up to tackling their own paddles. I must confess that I often buy my nopales prepeeled and even freshly chopped from Mexican groceries. When shopping, look for nopales that are bright green, small, and tender, and avoid any that are flabby and soft.

In addition to the slow-release foods and the nopales, many of the spices in Taurino's garden and kitchen have antidiabetic properties. This is not surprising, given that some of the pharmaceuticals that are used to treat diabetes (such as Glucophage) are derived from plants. In general, herbs alone are not adequate treatment for someone who already has a diagnosis of type 2 diabetes. Pharmaceuticals are also needed. However, regardless of the severity of your diabetes, adding herbs and other slow-release foods to your diet (plus exercise) can help you lower your dose of medication.

## Slow Herbs and Spices That Help You Control Your Blood Sugar

- Cinnamon
- Cloves
- Fenugreek seeds
- Parsley
- Garlic
- Cumin seeds
- Ginger
- Mustard leaves and seed
- Curry leaves
- Coriander seeds

## Arturo's Hesitations

When I first gave Arturo some Tarahumara recipes, he voiced three main concerns. First of all, wasn't a high-protein, low-carbohydrate diet better for preventing diabetes? Far from being low in carbohydrate, these recipes were relatively "high carb." Second, was this diet nutritionally sound? Would it control his blood sugar but, in the end, make him deficient in essential nutrients? And finally, what about the fat? Arturo noticed that some of the recipes called for lard (or butter). He was under the impression that *lard* was a four-letter word when it came to nutritional health—especially for diabetics. How could this be a good addition to any diet? These were all very reasonable questions. And while there is quite a bit of debate about which diet is best for diabetes, here are some of the facts.

## High Protein versus High Carb?

When high-protein, low-carbohydrate diets are compared calorie for calorie with diets that feature a fair amount of slow-release carbohydrates *both diets seem to have a similar effect on controlling blood sugar.* Let's review the advantages and disadvantages of these two eating options. According to a couple of studies, high-protein, low-carbohydrate diets help obese people with high insulin levels lose weight a little faster. (One recent study had diabetics getting as little as 20 percent of their calories from carbohydrates!) Since diabetes is caused or exacerbated by body fat, any diet that promotes weight loss can be considered beneficial. Unfortunately, I have several major concerns about cutting carbohydrates down to a bare minimum and replacing them with big servings of protein. The first of these concerns has to do with staying power. As Arturo had already discovered, most people who cut carbs out of their diets eventually find themselves craving starchy foods at every meal. In the end, this often results in overcompensating and binging on these foods. My second concern has to do with the effect that protein itself can have on our systems. Protein is not a very clean-burning fuel, producing ammonia by-products that can be harmful to the kidneys. While this may not be a significant matter for most people, it is a real concern for people with diabetes, prediabetes, and/or hypertension whose kidney function is often compromised by their chronic health issues.

By contrast, most diets enjoyed in diabetes cold spots around the world are packed with slow-release carbs. One prominent example is the indigenous Hawaiian Waianae diet that has been adopted by many diabetics and prediabetics in Kauila Clark's community. Like the Tarahumaran diet, over 70 percent of total calories in this diet are from slow-release carbohydrates, and yet it has been shown to lower blood sugar and even reverse diabetes. In addition, most people find this way of eating to be satisfying and easy to stick with for the long run. When I weigh all the evidence, I tend to lean more toward modest serving sizes of indigenous foods like the ones discussed in this book. Most of these diets include at least 60 percent of their calories from traditional slow-release carbohydrates.

## What About Nutritional Values?

Apparently the Connor team also shared Arturo's concerns about the nutritional soundness of the Tarahumara diet since they decided to analyze a sample meal. To their surprise, while the Tarahumara diet did not mirror the USDA recommended guidelines, it was nutritionally sound. The calories were consistent with the standard recommendation for adults in the United States with a similar activity level. Furthermore, protein and fiber were plentiful and a typical meal seemed to have enough of valuable micronutrients such as calcium, vitamins A and D, folic acid, and vitamin $B_{12}$.

## What About the Lard?

When William Connor spent time with the Tarahumaras, he was surprised at how little additional fat was used in their cooking. Vegetable oils (which now sometimes make it to the ranchos) were only available in distant stores, required cash, and were heavy to lug for miles on foot. Most Tarahumaras preferred not to eat their precious animals (except on special occasions), making animal fat hard to come by. In fact, Maria Cerqueira, a nutritionist working with the Tarahumara, calculated that less than 10 percent of their daily calories came from animal fats and vegetable oils. This stands in sharp contrast to the average American diet, where over 30 percent of the calories come from these sources. Despite the fact that they have very little added fat in their diet, most Tarahumaras seem to get a fair amount of healthy omega-3 fats from sources such as wild greens, nuts, seeds, seasonal fish, free-range eggs, and wild game. They also get an abundance of cholesterol-lowering fats (called sterols) from beans and corn. Diets rich in omega-3 fats and sterols have been shown to be protective against diabetes as well as other modern chronic diseases.

On the occasions when added fat is used, it is usually a small teaspoonful of pork or other animal fat. Far from being a four-letter word, lard is a better nutritional choice than many processed cooking oils. First of all, a little goes a long way in terms of giving foods flavor and a creamy consistency. In addition, if the pig was sharing its owner's omega-3–rich diet, then the lard itself will be an excellent source of these fats. Lard certainly offers less saturated fat than the same amount of butter and less inflammatory omega-6 fats than most vegetable oils. Dairy fats, which are an enormous source of saturated fat and calories in our Western diet, are consumed only in small amounts by the Tarahumara. Large-scale cattle or goat farming is generally required to produce milk in any quantity, and these milk products need electricity for refrigeration, something which is not available in most Tarahumara homes. Occasionally sheep or goat milk is fermented into a soft cheese and then used as a garnish atop stews and other dishes. Far from having a big impact on blood sugar, when fermented dairy (as well as other fermented edibles) are eaten in this manner, they can actually help control the breakdown of the carbohydrates in the meal, thus making the entire dish into a more "slow-release" food.

## Lard Reconsidered

In small amounts, lard gives tremendous flavor to a batch of beans and actually has less saturated fat than the same amount of butter. A little goes a long way. Use 1 tablespoon for every 3 cups of dried beans. (One tablespoon of lard has 115 calories, 13 grams of fat, 5 grams of saturated fat.) Render your own lard by cooking bacon at a low heat and skimming off the fat. Or you can buy pure, unpreserved lard prepared from certified organic hogs. Steer clear of lard that has added hydrogenated (trans) fats or preservatives.

## Back to the Burrito

At some point during my travels through Copper Canyon, I remembered Arturo's burrito. Was there any part of that meal that I had eaten in the corner taqueria in San Francisco that could be considered authentic Mexican food? It did not take much digging to find out. In a small village near Creel, I wandered into a luncheon shack. Looking everywhere for a menu, I finally gave up and asked the woman behind the counter about my options. She smiled, shrugged, and answered, "*Burritos, nada mas*" ("Burritos, nothing more"). Excited, I ordered four. At last I would meet a traditional burrito, made in its homeland.

After several minutes, she handed me four small, warm rolls covered in paper. Each one was no larger than a big cigar, six inches long and one and a half inches in diameter. Nothing like Arturo's massive burrito! I opened one and discovered a handmade corn tortilla stuffed with a half cup of black beans, a layer of red chili sauce, and some tender shredded meat. When I asked the server what kind of meat, she answered "*burro* [donkey]." I tried unsuccessfully to disguise my shock, and she laughed, telling me that she was not being serious—or was she? After all, that would finally explain the name. I wandered out into the sunshine holding my little package and marveling at how this simple food could have morphed into the complex, calorie-crammed San Francisco burrito that Arturo enjoyed on a nightly basis.

| | Calories | Fat (g) | Sat Fat (g) | Carb (g) | Protein (g) | Sodium (mg) |
|---|---|---|---|---|---|---|
| Arturo's burrito | 1,174 | 60 | 23 | 142 | 52 | 2,295 |
| Tarahumara burrito | 215 | 5 | 0.6 | 14 | 29 | 170 |

## Eating Like a Tarahumara

*Arturo began shopping in some of the Mexican food stores in San Francisco's Mission District and frequenting a local natural food market. Over the next six months, his weight slowly dropped (approximately one to two pounds per month), and his fasting blood sugars came back to normal. Even his blood pressure, which was high normal to start with, came down into the truly normal range. Meanwhile, he increased his trips to the gym to four to five times a week, finding it easier to exercise as he lost weight. His insistence that any recipe he attempted should be delicious yet simple had served him well—finally he had found a diet that he could stick with for the long term.*

*The last time I saw Arturo he was back to his joking self. He told me that he still went to his favorite taqueria once a week, but now he made the rest of his meals at home. Recently, he had met a new girlfriend who liked Harleys, had a good sense of humor, and enjoyed tasty food. Her only fault, it seemed, was that she was not a very good cook. "But that's okay," he said. "These days I do the cooking."*

## Three Steps to Controlling Blood Sugar the Tarahumara Way

### Step 1 (Basic):

- Think native grains. Stock up on the three sisters—corn, squash, and beans.
- Skip processed grains, such as white-flour pastas and breads and white rice, and processed legumes such as refried canned beans.
- Veggies and more veggies! Accompany indigenous food staples with plenty of vegetables: peppers, nopales, onions, cilantro, tomatoes.
- Make meat a spice: If you eat meat, use it as one flavor in the meal and not as the main part of the meal.
- Replace sodas, juices, and other sweetened drinks with water or unsweetened tea.
- Control your portion size and eat slowly to prevent a spike in glucose.
- Walk like a Tarahumara! Do some form of aerobic exercise for at least thirty minutes, four days a week.
- Cook at least one recipe per week from the Copper Canyon recipe section.

### Add Step 2 (Intermediate):

- Good fats: Get your fats from a variety of nonmeat and nondairy sources, including nuts, seeds, fish, avocado, olives, and whole grains.
- Tips for eating out: Order half-size portions or share with a friend. For Latin food, hold the nontraditional additions (like sour cream, loads of cheese, or the white flour tortilla), and stick with the beans, veggies, chicken, and two small corn tortillas.

- Mix your meals. Combining fats, proteins, and slow-release carbohydrates slows down the absorption of sugar into the blood stream.
- Don't overcook your starches. Slightly undercooked grains (al dente) produce a smaller rise in blood glucose than mushy ones.
- Slow down those carbs. Add fermented foods such as vinegar and yogurt to your meals as these slow the release of sugar from your carbohydrates.
- Consider using a small amount (1 teaspoon per serving) of lard for flavor rather than larger amounts of commercial vegetable oil.
- Do some form of aerobic exercise for at least thirty minutes, five days a week—consider hiking like the Tarahumaras.
- Cook at least two recipes per week from the Copper Canyon recipe section.

### Add Step 3 (Advanced):
- Buy handmade corn or wheat tortillas—or try making your own. Eat them with a bean filling, and this will make the whole meal more slow release.
- Cook with generous amounts of antidiabetic herbs and spices.
- Be adventurous. Eat nopal cactus once a week. You can buy them precooked in a jar or make the delicious recipe on page 256.
- Reserve desserts for feasts and special occasions.
- Exercise for at least thirty minutes, six days a week.
- Cook at least three recipes per week from the Copper Canyon recipe section.

# 6. Crete, Greece:
## A Cold Spot for
## Heart Disease

This voyage takes me to an island that has gained international recognition as cold spot for heart disease. Heart disease, or cardiovascular disease, is now the number one killer in the United States and, like other chronic diseases discussed in this book, it seems to increase in perfect lockstep with the speed of industrialization in any given region. Therefore, thanks to globalization, by 2010, the World Health Organization fully expects that cardiovascular problems will replace infections and famine as the greatest cause of death and disability worldwide.

Heart disease is a general term used to describe a whole host of health issues related to the circulatory system. These include high cholesterol, high blood pressure, atherosclerosis, and, in the most extreme cases, heart attack or stroke. Most people I know have been personally affected or have a close family member who has been affected by at least one of these problems. In addition, heart disease and diabetes often coexist since they both share many of the same risk factors including obesity, a sedentary lifestyle, and a suboptimal diet. A person who is beginning to show signs of prediabetes plus heart disease is considered to have "metabolic syndrome," or "Syndrome X." This syndrome increases one's lifetime risk of heart attack or a stroke by almost 400 percent.

## Do I Have Metabolic Syndrome?

The following are the most widely accepted criteria for Metabolic Syndrome:

A waist circumference greater than or equal to 40 inches (for men) or 35 inches (for women), *plus* any two of the following:

- Blood pressure greater than or equal to 135/85, or you are already taking blood pressure medication
- Blood triglycerides greater than or equal to 150 mg/dl
- Fasting blood sugar greater than or equal to 100 mg/dl
- HDL cholesterol less than 40 mg/dl in men or 50 mg/dl in women

In this chapter, I introduce Tanya, a woman with metabolic syndrome and a family history of heart disease, who is seeking a diet that will help restore her health. Inspired by Tanya, I learn about the Seven Countries Study, a landmark study that established the island of Crete as a heart disease cold spot in the Mediterranean region. This is *the* cold spot that nutritionists and other health experts are referring to (whether they know it or not) when they sing the praises of the Mediterranean diet. Before I travel to Crete, a cardiology professor at Harvard offers his perspective on the root causes of heart disease and metabolic syndrome and the relative value of food versus medicine for preventing and treating these health problems. Next, I make my way to Crete where a food activist, a physician, a farmer, an organic grocer, a chef and wild food hunter, a waiter, and a small-town entrepreneur all offer their opinions about the most powerful heart-healthy foods. Finally, an Athens researcher helps me put it all together, and the true healing nature of this diet is revealed.

## Meet Tanya

Tanya first came to see me several days after her thirty-eighth birthday. Her story was similar to ones that I had heard often enough. She lived in San Francisco with her husband and two small children but had grown up in New Jersey where her parents owned and operated a family-style diner near the interstate. Prior to having her kids, she considered herself healthy and fit. However, after each pregnancy, she had gained an additional twenty pounds and now felt frustrated by the fact that she was forty pounds overweight. She also noticed that with each doctor's visit, her blood pressure seemed to creep up another couple of points. This confused her since she felt that she ate the "healthiest diet in the world," the Mediterranean diet.

Both of Tanya's parents, who still live in the Northeast, had been battling obesity and high blood pressure since their early forties. Her father had recently had a serious heart attack and was taking what Tanya described as a "slew" of medications. Meanwhile, his parents still lived in their little village in northern Italy and were vigorous and free of heart disease even in their late eighties. Tanya wanted to know what she could do to be more like her grandparents and not follow in her father's footsteps. She wanted to lose weight, reverse her upward-creeping blood pressure, and ward off a heart attack.

To help Tanya meet her goals, I asked her to write down a list of all the foods she had eaten over the course of a week. I figured that a food diary would help me to better understand the contents of her Mediterranean diet. I also gave her a lab slip and asked her to get some blood work done. I was especially interested in her cholesterol and blood sugar levels.

When Tanya came back for her second visit, we reviewed her lab results. Her fasting blood sugar was still within the normal range, but her total cholesterol and fat (triglyceride) levels were both higher than normal, and her HDL-C (the high-density, "good" cholesterol) was low. In addition, a newer test for heart disease called "C-reactive protein" indicated that Tanya was at higher than average risk for cardiovascular problems. Given her parents' heart troubles and her borderline high blood pressure, I found these numbers worrisome. Tanya clearly shared my concerns about her heart health. Nonetheless, she was insistent that we try alternative approaches before starting her on medicines for blood pressure and cholesterol control. After all, she was only thirty-eight and did not want to be popping pills daily for the rest of her life.

While Tanya's request seemed like a reasonable one, it suddenly became doubly important that I discover the contents of her Mediterranean diet. She produced her food log nicely organized by date and food categories. The most commonly listed foods included 1) pizza, 2) pasta with tomato sauce and Parmesan cheese, 3) white ciabatta garlic bread dipped in olive oil, and 4) cheese ravioli. These foods, which she had learned to love as a child while eating at her parents' diner, had remained the mainstay of her diet.

## In Search of the *Real* Mediterranean Diet

Technically Tanya was right—she was eating a Mediterranean-inspired diet. But were these meals, at least in the form that she was eating them, a part of *the* Mediterranean diet that has been praised far and wide for its role in preventing heart disease? Tanya's lab results suggested otherwise. Indeed, she was not the only patient in my practice with heart issues who claimed pizza as her favorite food. I wanted to know *which* Mediterranean foods were responsible for making this diet a household word. After all, the Mediterranean is a big sea surrounded by sixteen countries and encompassing the two island nations of Malta and Cyprus. There are huge variations in food choices and recipes between one country and the next. For example, in southern Italy, goat and sheep cheese are part of the staple diet, and broccoli, zucchini, and green beans accompany many meals. In Morocco, where I lived for several years, cheese is rarely found in traditional cuisine, and I remember many exotic spices but nary a broccoli head. Fish is plentiful in Spanish, Greek, and Italian food but prepared much less often in Bosnia and Croatia, where they seem to prefer red meat. In these Slavic countries, potatoes and pumpkins are a staple, and they often use heavy cream as a garnish. In Greece and Lebanon, however, dried beans are one of the main staples, and yogurt replaces the heavy cream. Where, among all these countries, was the Mediterranean diet that was truly deserving of being called "heart healthy"?

To answer this question, I turned to the writings of the late Ancel Keys, PhD, a man who is widely recognized as one of the grandfathers of modern epidemiological research. Keys was a physiologist at the University of Minnesota where he spent the early part of his career doing nutrition research on K ration packages for the U.S. military. One of the perks of this job was that he got to travel a fair amount around Europe. During these voyages, he found himself particularly enchanted by the countries surrounding the Mediterranean; he loved the pace of life, the weather, and of course, the food.

Ancel Keys was especially impressed by the excellent health enjoyed by the denizens of these countries. While the rate of heart disease among middle-aged men in his own country was skyrocketing, men in countries such as Italy and Greece were living long, heart disease–free lives. This led him to pose the very reasonable question: "If some developed countries can do without heart attacks, why can't we?"

In 1958, Ancel Keys began a landmark study called the Seven Countries Study to solve this mystery. He decided to look specifically at the life expectancies of 12,000 middle-aged men living in Italy, the Greek Islands (Corfu and Crete), Yugoslavia, the Netherlands, Finland, Japan, and his hometown of Minneapolis, Minnesota. His goal was to determine which group lived the longest and why. Forty years later, when the research team paid a follow-up visit to all the study participants, they discovered that the Italian and Greek men were more likely to have survived. These men had also suffered relatively fewer heart attacks and strokes. Focusing within this group, the healthiest of all were the 686 men living in eleven villages on the island of Crete. Their survival rate was almost twice as high as the men from Minneapolis, and, not surprisingly, they had experienced much less heart disease. These numbers spoke for themselves. There was something that they were doing just right on Crete.

But what was this *something*? The Seven Countries Study research team had collected detailed information from all the study participants. They had determined whether or not they smoked, what they did for a living, and how much they exercised. After a lengthy analysis, the only factor that truly set the men from Crete apart from the other men in the study was their diet. These findings, which are now published in many languages, are what initially brought the heart-healthy benefits of the Mediterranean diet to international attention.

In the years that followed, the Seven Countries Study spawned dozens of others around the globe. In each instance, the goal was to see whether variations of the diet from Crete would prevent or reverse heart disease. Slowly the results of these trials rolled in. Communities as far flung as Oslo, Lyon, Moradabad (India), and Los Angeles were reporting success at warding off first heart attacks or lowering the likelihood of a second one by eating Cretan-style. So there you have it. When nutritionists, physicians, nurses, and researchers talk about a Mediterranean diet that is proven to prevent or treat heart disease, they are usually referring to traditional Cretan food.

## A Cardiologist Gives His Blessing

Despite the promising results from all these nutrition studies, it is hard not to be influenced by the endless procession of ads for cholesterol-lowering medicine and blood pressure pills that one sees on television and in magazines. Such messages would have you believe that drugs are indispensable for preventing or treating heart disease. Is this truly the case, or can the right foods actually replace drugs? I wanted the perspective of an expert, so I called Paul M. Ridker, MD, who is the Eugene Braunwald Professor of Medicine at Harvard Medical School and director of the Center for Cardiovascular Disease Prevention.

"When it comes to preventing or treating heart disease," explained Ridker, "the important thing is to select treatments that are anti-inflammatory, anticlotting, and that slow the release of sugar into the bloodstream. These treatments can be either a drug or a diet like the Mediterranean diet." He then elaborated on his answer: "From early history, most people died from trauma, starvation, or infection. As a result of these threats, our bodies have evolved to do three things very well: clot, hold onto calories from foods, and give a strong inflammatory immune response to any foreign substance." As he sees it, these protective mechanisms served us well in the premodern world. However, in modern society traumatic events are so rare that they make headline news, high-calorie food is cheap and abundant, and antibiotics and vaccines have eliminated the threat of most serious infections. Therefore, these natural defenses of clotting, calorie hoarding, and inflammation have become maladaptive or dysfunctional and have led to cardiovascular disease as well as other chronic diseases.

"There is good news, however," continued Paul Ridker. "This process is easy to reverse." He explained that both diet and drugs can be effective treatments since either one can lower the amount of inflammation and clotting in the arteries. In fact, in his opinion, drugs like Lipitor probably have their most powerful effect not by lowering cholesterol levels but by preventing inflammation, acting much in the same manner as foods rich in omega-3 fats or antioxidants. (As an aside, this explains why Ridker believes that blood levels of C-reactive protein—a protein which increases when inflammation is present—might offer a better test for heart disease than the standard cholesterol panels.)

Studying the medical literature, I found a good deal of support for Paul Ridker's comments. When cholesterol-lowering medications were compared head-to-head with a diet like the Mediterranean diet (plus thirty minutes of exercise four to five days a week), they both had similar benefits in terms of preventing heart disease. The only difference was that statins, the drugs most commonly prescribed to lower cholesterol, can be swallowed once a day with a glass of water, while diet therapy and exercise require commitment and effort to have an effect. While it might be very enticing to take a pill and forget the hard work, I believe that there are several very good arguments for choosing the diet and exercise route. First of all, there are a number of potential side effects associated with statin drugs, and their long-term consequences are still not fully understood. Second, dietary change and exercise offer benefits that go far beyond heart disease prevention; they improve your sense of well-being and decrease your risk of developing other chronic diseases. When I asked Ridker which path he would choose, he told me he would first take all the necessary steps to improve his eating patterns and lifestyle and then, if necessary, would move on to drugs. Bolstered by these comments, as well as my own experiences in my medical practice, I resolved to learn what I could about the Cretan heart disease prevention diet.

## When to Use Drugs for Lowering Cholesterol

In my practice, I reserve medication for people with high cholesterol and/or C-reactive protein who have

- made a good attempt at exercise and changing their diets and *still* have abnormal labs,
- no interest in changing their diet, or
- other serious risk factors for heart disease such as a previous heart attack, diabetes, high blood pressure, or a very strong family history of early heart attack or stroke.

Even in these situations, I encourage maximal effort at diet and life-style change. This approach improves your overall health and sense of well-being. It also increases the effectiveness of your medication and could help you to lower your daily dose.

## Can You Treat High Blood Pressure with Diet?

There is a fair amount of evidence that a heart-healthy low-salt diet and exercise should also be one of the first-line therapies for high blood pressure.

If you are diagnosed with prehypertension (systolic pressure between 120 and 129 mm Hg or a diastolic pressure between 80 and 89 mm Hg) then it usually makes sense to treat yourself using diet and exercise.

If you are diagnosed with hypertension (systolic pressure higher than 139 mm Hg or diastolic pressure higher than 89 mm Hg) then it is advisable to start medication *and* treat yourself using diet and exercise.

## The Food Activist

It was two weeks before Greek Easter when I found myself on the northeastern coast of Crete, driving along in my minuscule rented car. I was starting my odyssey to discover the real Mediterranean diet with a visit to chef and cultural activist Nikki Rose. Having spent some of my plane ride reading excerpts from Theocharis Detorakis's *History of Crete*, it had occurred to me that it was not just by chance that Crete emerged as the ultimate cold spot in the Seven Countries Study. After all, Crete was home base to the Minoans who, according to Detorakis, had the distinction of being one of the first advanced civilizations in Europe. Since 3400 BC, the residents of this island had been borrowing the best seeds and farming practices from surrounding countries and, in turn, exporting edible goods throughout the Mediterranean basin and beyond.

Leaving the airport near Heraklion, I wondered what those Minoans would make of the scenery that now lined the main highway. For miles, all I could see were sprawling concrete vacation resorts, each one a separate testament to the ill effect of mass tourism. Periodically, I would pass a large roadside restaurant, its parking lot filled with tour buses, with a neon sign advertising "Real Cretan Food." I seriously doubted this claim.

Just as I was beginning to despair of seeing the real Crete, the highway gave way to a bucolic, winding road, and the resorts disappeared. Slowing down, I opened the car window to let in the sea breeze and the springtime sun, a sun which was already intense enough to give my bones a good warming. Lemon and orange trees were in blossom, and the sweet fragrance filled the car. Each twist in the road gave me new dramatic views of crimson seaside cliffs, emerald green fields of wild grass, and an aquamarine

Aegean Sea. Occasionally, on a stretch of road between the white-washed hamlets, I would pass a clump of wild purple irises or red poppies. Despite the risk of a rear-ender, I found myself involuntarily jamming on the breaks to feast on the view. Everywhere were herds of sheep and goats, their latest offspring fleeing from my car fender as they desperately tried to keep up with the elders. Stopping in a roadside taverna, I sipped a glass of freshly squeezed orange juice—nectar of the gods—and considered my good fortune. In addition to being a promising cold spot for heart disease, this island was turning out to be an ideal vacation locale.

By early afternoon, I had made my way to Nikki Rose's seaside apartment. "Just in time for meze!" were her first words as I pulled into the driveway. Nikki was hard at work tilling the earth in a small garden plot that lay right next to her apartment. With my arrival, she brushed the dirt from her knees and led me over to a nearby taverna. We sat out on the terrace overlooking the ocean, and, as I admired the view, Nikki ordered in Greek. "I just asked for whatever is good and fresh," she said.

## What Is a Meze?

Meze are not technically appetizers. Similar to tapas in Spain or *okazu* in Japan, they are small dishes that collectively make up a meal and are shared by everyone at the table.

As we enjoyed succulent octopus and tender fresh fava beans, Nikki told me that her mother was born and raised in the highlands of Anogia, a part of Crete known for its rebels and independent thinkers. Nikki herself grew up in Washington, D.C., but I could easily see how this small, intense woman with a no-nonsense way of speaking had retained a fair amount of the Anogia spirit. As a trained chef, she started visiting Crete in the 1990s to sample her ancestral foods and learn the recipes. However, she swiftly realized that there was another job to be done: She needed to preserve these food traditions before they disappeared forever. With the death of each grandmother and the sale of each family farm, she saw recipes being lost and organic farming practices becoming obsolete. To help preserve Crete's culinary heritage, Nikki eventually made the bold decision to move to Crete full time to found Cookingcrete.com. In a relatively short time frame, Nikki has made it her business to know everything there is to know about Crete and its food. Her one-woman operation has forged alliances between organic growers and buyers and has helped promote farms and local chefs by featuring them in culinary tours and on her website.

"Mind you, it's an uphill battle," sighed Nikki, gazing out at the sea. "Many Greeks in the cities do not understand what a treasure they have on this island." Recently she had pitched a story about traditional cheese making on Crete to a television station in Athens. "The producers were not impressed. They said, 'Why do we want to watch some country bumpkin making local cheese?' This was depressing." After a pause, however, her voice became more hopeful. "Maybe your readers will get it."

With this she pulled out a detailed map of the island and flattened it out on the table. As she told me about each one of her favorite food entrepreneurs, be they farmers, bakers, vintners, chefs, or grocers, she drew a little star to mark their whereabouts. By the time we were done with our three-hour lunch, I had an itinerary that I felt certain would help me uncover all the healing secrets of the Cretan diet.

## Another Warming Cold Spot

En route to my first point of interest on Nikki's map, I made a side trip to meet with Artemis Pontikaki, MD, a young physician working in a rural clinic in eastern Crete. Walking into the main waiting area of the clinic, I was greeted by a thick cloud of tobacco smoke. Through the haze I noticed a couple of women dressed in nurse's whites who were standing in the corner and drinking coffee while puffing on an early morning cigarette. Behind them were two "no smoking" posters featuring anatomical drawings of unhealthy lungs and hearts. Talk about mixed messages!

In this clinical setting, Artemis, wearing blue eye shadow, tight jeans, stiletto heels, and a cropped shirt, seemed almost as out of place as the tobacco smoke. After offering me some Greek coffee as thick as mud, she told me that she had grown up in the more cosmopolitan area of Hania but was working in this small town to fulfill her National Health Service requirement. During the course of our conversation, I got the impression that Artemis was bored with her job. There was not very much for a city girl to do in this rural post, and the work was very slow. Apparently the empty waiting room was a fairly typical sight. I asked her how many heart attacks she had seen over the six months that she had been working in the region. She thought for a while and then answered "two." Both were in people over the age of eighty. Indeed, for a young doctor fresh out of training, it could get boring to care for a group of people who were so darn healthy.

Artemis then mentioned that times were changing. Most of her younger friends and physician colleagues smoked, and they preferred fast food to the recipes that were served up in their grandmothers' kitchens. As a result, while the older folks were still going strong, she was seeing a surge in high blood pressure and heart disease in people in their thirties, forties, and fifties. When I asked her what could be done about this, she shrugged and said, "We should spend more time with people like my grandparents in Hania and learn why they are so healthy."

After leaving the clinic, I headed out on back roads, making my way westward across the island. I could not deny that the smoky clinic and Artemis's parting observations left me feeling a little depressed. Like Copper Canyon, Crete was a warming cold spot. However, I soon found myself driving through more tiny villages where I saw dozens and dozens of healthy-looking seniors all going about their daily business. Some of the older folks were gardening, others were tending animals or repairing houses, and many more were sitting and socializing in the noontime sun outside the village tavernas. In one town center, I had to slow down to make room for a village elder who, midroadway, appeared to be having a wrestling match with her two goats. With each encounter I was greeted with a wave and a friendly "Yiasas!" which in Greek means "to your health." Thanks to these scenes, I approached my mission with renewed resolve. In the week to follow, I needed to learn why people like Artemis's grandparents had stayed so healthy.

My next destination was the town of Astrikas on the western side of the island. There I planned to visit Nikki's friend Yiorgos, owner of the Biolea olive and olive-oil press. As I drove into Astrikas, I stopped along the road and asked an older man for directions to Yiorgos's farm. Looking confused, he asked, "Which Yiorgos?" Foolishly, I did not realize that Yiorgos (aka George) is one of the most popular names in Crete; there are probably thirty or more Yiorgoses just in tiny Astrikas. I dug out my map and produced the last name Dimitriadis. "Aha!" he said. Crossing the road, he picked a small branch of a nearby olive tree and handed it to me as an offering. Yes indeed! I was looking for Yiorgos, the olive man. My informant jumped in his truck and led me up the hill.

Amidst the acres of olive trees, I parked in front of an imposing new Tuscan-style building complete with thick walls, high wooden trusses, and stone archways. Yiorgos Dimitriadis, a handsome, fit man in his mid-fifties with graying temples and a coal-black mustache, was standing at the entrance wearing a white lab coat. He was in the process of overseeing workers as they loaded several dozen cases of his artisanal olive oil—each one bound for the shelves of Whole Foods and Williams-Sonoma in the United States.

Yiorgos is a man with strong opinions—especially when it comes to olive oil—and he is not above using some well-chosen English or Greek expletives to get his point across. According to him, this oil is one of the main secrets to Cretan longevity and heart health. Take his ancestors, for example. They have farmed, pressed, and eaten olives for at least six generations (and probably for hundreds, if not thousands, of years before that), and their health was exemplary. Proudly, he showed me a photograph of his great-grandparents that had been scanned onto the screensaver of his office computer. Collectively the couple had consumed over fifty kilos of olive oil per year, a fact which Yiorgos believed helped them both live well beyond their hundredth birthdays. His grandfather put away similar quantities of the stuff and probably would have also made it to a hundred if he had not died at eighty-five after falling off a construction ladder.

Yiorgos's personal health story, however, had taken a very different course. As we stood together on the hill surveying his thirty hectares of neatly spaced olive trees, he told me about how he had strayed from his olive-rich heritage. Although he grew up in Astrikas and had spent his childhood years on the farm, he made the decision in his early twenties to move to Canada and become a pilot. In North America, Yiorgos adopted a radically different lifestyle than the one he had led on Crete. "I left here. We hardly had meat once a week. I go to Canada. Beautiful! Breakfast meat, lunch meat, dinner meat! I also started smoking." When he was forty, Yiorgos returned to Astrikas to visit his family and take a vacation. While jogging along his favorite hill trails, he noticed that he felt winded and tired but attributed these symptoms to smoking and lack of regular exercise. Just to be on the safe side, however, he went to see his family doctor on Crete. As it turns out, it was a good thing that he did this. After undergoing some tests, Yiorgos was told that four of the main vessels to his heart were almost completely blocked. Days later, he was in the operating room for a quadruple coronary bypass.

There is nothing like a serious diagnosis and a massive operation to inspire someone to make a lifestyle change. Within months, Yiorgos had quit the high-stress North American lifestyle and moved his wife and kids back to Astrikas to begin growing olives. During his first season, he decided he would do everything possible to get a big crop. To this end, he used a wide range of conventional growing methods, including pesticides, fertilizers, and modern presses. However, at the end of the season, he was not satisfied with the product. He had sent a sample of the olive oil to a lab for testing and was dismayed by the report: There was still a significant amount of pesticide in his precious oil. In addition, the modern techniques for pressing the olives used large quantities of water. Since most antioxidants are water soluble, he had washed out many of the powerful protective ingredients from his oil. Finally, he tested a sample from the family well and discovered that a high concentration of the fertilizers and pesticides were also making their way into this water. By the next season, Yiorgos had changed everything. He had converted over to organic farming systems and was using a stone press for the olive oil. These were exactly the same techniques that his ancestors had used for centuries—maybe even as far back as the Minoans.

## What's So Good About Olive Oil?

I followed Yiorgos into the building and admired the three huge crushing stones sitting in a spotless stainless-steel bowl. Nearby were enormous vats filled with the honey-colored oil. My mind wandered back to the handsome ancestors on the screen saver, and I thought about all the olive oil that they consumed. Was this oil really one of the secrets to cardiovascular health and general well-being? And if so, why? The scanty nutrition education that I had received in medical school had taught me that monounsaturated fat or oleic acid in olive oil acted like a sponge to suck up "bad" cholesterol. However, the more I learned about nutrition, the less sense this made. Monounsaturated fat is found in abundance in poultry, pork, and a variety of vegetable oils, and the percentage of monounsaturated fat consumed by people eating Mediterranean style is only slightly more than other Western diets. Therefore, there must be something in the olive oil, other than the type of fatty acid, that makes it so good for the heart.

## The Free Radical Fighter with a Bite

Yiorgos handed me a tiny cup filled with olive oil so that I could sample the product of his latest press. With my first sip, I was suddenly reminded of my last visit to an olive oil farm. Almost a year earlier, my nutritionist friend Allison Fragakis and I spent an afternoon at the McEvoy Olive Oil Ranch located one hour north of San Francisco. While this farm is practically halfway around the globe from Astrikas, the landscapes of the two places are so similar that one would be hard put to tell the difference between the two locales. We were greeted at McEvoy by Jill Lee, the ranch hostess, who happens to have a master's degree in nutrition and who seemed to know everything there was to know about nutrition and olive oil. After a tour of the farm, Jill sat down with us at a big wooden table and ceremoniously broke the seal on a tiny glass jar filled with a golden green liquid. This was *olio nuovo* from the previous fall that had been pressed in a stone mill very similar to the one at Biolea. She poured us each a table-spoonful in a shot glass and asked us to give it a sniff. It smelled like lemon and fresh-mown grass with a hint of lavender. Then she completely covered her glass with both hands, as if it were a small animal that needed warming. We followed suit. Finally, she dipped the tip of her tongue into the oil and breathed in quickly and deeply. I mimicked this ritual, experiencing the most interesting series of taste sensations: first the tip of my tongue registered complete sweetness, almost like sugar candy. Then the sides of my tongue started to vibrate and tingle. Finally I felt a fresh grassy taste in the back of my throat. This pungency, Jill explained, was at least partially caused by polyphenols, which are strong antioxidants, in the olive oil.

Recalling my McEvoy experience, I dipped my tongue into Yiorgos's product and took in several quick breaths. Sure enough, there was that perfect balance of sweetness in my nose and mouth with the faintest of tingles in the back of my throat. Ah, these same antioxidants seemed to be residing in the olives of Crete. This cluster of sensory experiences represented *real* medicine; the same medicine that cardiologist Paul Ridker felt was responsible for preventing or reversing heart disease.

## Olive Oil: Medicine Indeed

In one study, different grades of olive oil (from low-phenol content to high-phenol content) were fed to people with high cholesterol to measure the effects that these oils might produce. Sure enough, the high-phenol olive oil had many benefits: preventing oxidation of cholesterol, increasing levels of healthy HDL cholesterol, and decreasing thromboxane B2 (a clotting factor). Not surprisingly, the low-phenol olive oil had none of these benefits. Yes, the polyphenols in olive oil seem to play a key role in heart health.

## Good Plate Fellows

Certainly the high-polyphenol content of olive oil can partially explain why it is so heart healthy. But Lluís Serra-Majem, PhD, a researcher at the University of Barcelona in Spain, offers some additional reasons why olive oil is so terrific for your health. After administering diet questionnaires to 1,600 adults in Spain, he discovered that people who ate a lot of olive oil were more likely to eat green vegetables, whole grains, and fish while people who steered away from olive oil were more likely to eat sweets, processed cereals, refined breads, and foods high in refined vegetable oils and animal fats. Hmm . . . what did this mean? Maybe olive oil just happened to be an innocent bystander in the success of the Cretan diet? Maybe it was just a food that was preferred by people who happened to make other healthy food choices.

This, in fact, is *not* Serra-Majem's conclusion. Instead, he believes that olive oil plays an important role in making other healthy foods more appealing. Try a plate of boiled wild greens without olive oil, and you will discover that it pales in comparison with eating these same greens in a small lemony pool of golden oil. The same goes for fish, tomatoes, and even freshly cooked bulgur. Apparently Yiorgos agrees. "Olive oil," he stated, "is there to enhance your food. It's that simple."

In addition to making foods more tasty, cooking with olive oil allows micronutrients such as lycopene in tomatoes and beta-carotene in carrots to be more easily absorbed by the body. This is important because lycopene from foods (but not from pills) plays a role in cardiovascular disease prevention. One epidemiological study found that while lycopene supplements had few benefits, the risk of heart disease was reduced by 30 percent in people who ate seven or more half-cup servings per week of tomato-based foods, most of which had been prepared with olive oil.

Finally, if you are using olive oil, you are less likely to be using an omega-6 rich vegetable oil such as corn oil. As we have already discussed, omega-6 fats are essential fats that are present in unhealthy amounts in our modern diet. They act as inflammatory agents in our cells and blood stream, and they also compete with our body's ability to synthesize and process the heart-healthy essential omega-3 fats. Conversely, *nonessential* omega-9 fats in olive oil do not compete with omega-3 synthesis. Therefore, olive oil could be having a positive effect simply by replacing other less healthy oils.

So there you have it; there are at least four ways that olive oil has earned its reputation as an important part of Crete's heart disease prevention diet: as an inflammation fighter, a vehicle for the absorption of other nutrients, a replacement for less optimal oils, and, finally, as an enhancer of other healthy foods. In the course of my visit with Yiorgos, he did mention all these benefits. Nonetheless, at the end of the day, he felt that the reason to use olive oil is very simple. Handing me a three-liter can of his latest pressing, he said, "People like to get all froufrou about olive oil. I hear people say, 'I can taste fresh apples, or chocolate, or wine, or dadada.' I say bullshit! Just taste the olive oil."

## Too Much of a Good Thing: A Cautionary Note

Olive oil is very caloric (approximately 120 kcal per tablespoon), and therefore, like all good foods, moderation is key. For generations, people on Crete have managed to stay trim while consuming gallons of this oil because of the hours they spent each day doing hard physical labor. However, like all parts of the world, mechanization and industrialization have replaced a lot of this work, and one now sees a fair amount of heart disease and obesity even in the rural parts of Crete. The recipes in this book give you some guidelines for how much oil to use per meal to achieve that balance of nourishing the heart while sparing the waistline.

## The Other Plate Fellows

Standing in the kitchen of their Taverna Panorama in the seaport village of Milatos, I asked Nikki's friends, Katerina Astroulaki and her husband, Nektarios Androulakis, to tell me about the foods that they thought were responsible for their ancestors' good health. In unison, they both pointed to the colorful display laid out in front of us on their wide wooden counter. These were all the plate fellows that Lluís Serra-Majem had mentioned in his olive oil studies, foods that Katerina and Nektarios were about to transform into mezes to be served that evening to their taverna guests.

I admired the neat piles of ingredients: quartered lemons, bunches of parsley and oregano still dusted with earth and dew from the garden, leeks, minced garlic, red onions, eggplant, slender cucumbers with tiny seeds, and pods of fresh fava beans. On the other end of the counter was a large brick of soft white cheese that appeared to be homemade. Near the vegetables, in a stainless-steel tray, were freshly caught octopus, sardines, anchovies, and dorado fish.

Nektarios explained that since it was still Lent, they were abstaining from rich foods such as sweets and red meat. Therefore, most of their meals were vegetarian or contained fish and other non–red meat proteins. Disappearing into his storeroom, he returned with a mesh bag containing his favorite Lenten food: land snails. He described how, after painstakingly hunting for each one, he needed to keep them in an aerated box for twenty days and feed them cornmeal to clean out their systems. "Today," he explained triumphantly, "they are finally ready for eating."

In the center of Katerina and Nektarios's kitchen counter, surrounded by the vegetables, the cheese, and the fish, I noticed a huge pile of bright green weeds. At first, I wondered why the waste from of a day's worth of gardening was lying on the counter. But then it suddenly dawned on me that I had come face to face with my first *horta*, the wild greens that were reputed to play such an important role in heart health. These particular ones, a type of wild chicory called stamnagathi, were prehistoric looking with jagged curling leaves and twisted stems. I had assumed that the practice of horta picking had disappeared long ago, but Katerina assured me that the tradition was alive and well. She took every opportunity to go on horta hunts, and she was intent on teaching her young boys to do the same. As a matter of fact, she planned to go out the following morning and offered to take me along.

## Fish and Horta: Two Powerful Anti-Inflammatory and Anticlotting Foods

Wild seafood and snails are terrific sources of EPA, a powerful omega-3 fat. Omega-3 fats have been shown to prevent clotting in the arteries. In addition, they help to stabilize the heart's electrical conduction system and prevent sudden death from cardiac arrhythmias.

Wild greens (or horta) are rich in LNA, another form of omega-3 fat. They are also full of antioxidants (including vitamin C, beta-carotene, folate, and quercetin) and, therefore, help lower inflammatory proteins associated with heart disease. One study found that wild greens had a higher antioxidant effect than tea or wine. Adding acidic foods such as lemon or vinegar to horta helps to increase the bio-availabity of these nutrients and greatly enhances the taste.

## Hunting for Horta, or the Dirty Fingernail Sign

Early the next morning, as the sun was just stretching across the sea and the nighttime dew had not yet dried, Katerina and I set out on our horta hunt. Armed with a sharp knife and a plastic bag, she led me to a nearby hillside. As we made our way there, she explained that this was not her most secret horta spot but still it was pretty good. I understood. After all, even though we had been introduced through Nikki, I was a stranger and a writer at that. Who knows, maybe I would divulge the location of her precious hunting ground in my book.

When we reached the area, Katerina bent over what seemed like a nondescript area of weeds and deftly dug the tip of her knife and her index finger into the ground. Within seconds, she had extracted an intact clump of the same stamnagathi that I had seen on the counter the previous evening. I tried to locate my own to imitate her, but all the greenery looked the same. She pointed one out for me, and I applied her knife and index finger technique, but the weed disintegrated in my hands. She laughed and encouraged me to keep at it.

As Katerina began to fill her bag, I crawled over the hillside, nose inches from the ground. Slowly, the various weeds underfoot began to differentiate themselves, and I recognized the stamnagathi as well as a dandelion-type green that was also fair game for that day's hunt. After several tries with the knife and the finger, I too figured out the picking technique: first sever the weed from its root base with the point of the knife, and then use the nail of the index finger to pop the horta out of the ground. (For days to follow, I would regard the dirt buried deep under my nail as a badge of honor.) So soothing was the distant tinkle of goat bells, the sun on my back, and the pattern of stooping and picking that, before I knew it, two hours had passed. It was time to go back to the taverna for lunch and to meet Katerina's boys who would be coming home from school. As we headed back, I noticed that the quadricep muscles in my legs were sore from squatting—what an enjoyable form of exercise this had been!

At the taverna, we handed our bags over to Ioanna, Katerina's mother-in-law. My bag, for all my efforts, was not nearly as full as Katerina's, and I felt a bit inadequate. Sensing my insecurity, Ioanna assured me that I had done well for a first-time hunter. However, something about her facial expression suggested that she was just being polite. Sure enough, minutes later I spied her donating half my collection to the chicken pen. Apparently my technique needed some refinement.

For lunch we ate the horta stewed in olive oil and garlic and anointed with freshly squeezed lemon. While I was transported by the delicious flavor, the doctor in me could not help but be mindful of the many healing qualities of this simple dish: First there was the anti-inflammatory combination of the olive oil and the horta. Next, there was the garlic, which has anticlotting and blood pressure–lowering effects in its own right. And finally, there was the lemon, a wonderful source of vitamin C, which helps increase the bioavailability of the antioxidants in the horta.

At a nearby table, I noticed two men watching a soccer game on television. At first glance, they could have been a couple of men anywhere on the globe, drinking beer as they cheered and berated their favorite team. However, on closer observation, I realized that instead of the standard pretzels or nachos, they too were snacking on olive-oily shoots of fresh horta. Definitely a healthier option!

## Wild Greens in Your Neighborhood

Wild greens can also be found in your neighborhood. If you're interested in learning about picking your own wild greens, you'll need to learn from an expert. There are several clubs, books, and Web sites to get you started (see Appendix G). In my opinion, the next best thing to hunting for greens is to buy the freshly picked ones at your local farmer's market. Look for dandelion, Swiss chard, kale, beet greens, mustard greens, stinging nettles, sorrel, borage, arugula, broccoli rabe, spinach, and purslane.

## Bread Man on a Mission

In the lovely Venetian-walled port city of Hania, I wandered into one of the city's only natural food stores. A salesman was bent over, arms elbow-deep in a burlap bag of dark flour as he talked in an animated voice with one of his customers. Surrounding him were more bags filled with grain in varying shades of yellow and brown. Examining the contents I recognized oats, bulgur, and barley. Nearby another bag contained a golden-yellow flour that exuded the familiar aroma of chickpeas. After selling the woman several pounds of ground barley, the salesman came over to introduce himself. His name was Stelios Michelakis, and this was his store. I told him about my quest, and he wasted no time in giving me his opinion as to why many of his fellow islanders have such healthy hearts. "These," he said pointing to his bags of grains, "are some of the most important parts of our diet. I just sold that lady some fresh ground whole barley, but I also explained to her how to make it into bread."

Stelios has a thick shock of white hair and twinkling eyes that are the same bottle-green color as the Mediterranean Sea. As he served other customers, he told me about how he had found his way into the natural foods business. Over the course of his fifty-plus years, he had reinvented himself many times and had a wide variety of careers. However, in the end, he had to follow his heart and open one of the first organic food stores on the island. "If your mind is a little changed," he explained, "then it is difficult to do other work." Stelios, like Yiorgos, has long-lived ancestors who inspired him to pursue his life's work. In his case, it was his great-grandfather nicknamed Xeros (which literally means "dry" in Greek but, as a figure of speech, is used to describe someone with big muscles). Apparently, Xeros was famous in his home region of Almerida in northwestern Crete because he was so strong. On one historic day when the river had overflowed its borders, he was seen carrying a donkey overhead across the water to safety. "These days we are not as strong," Stelios laments. "For this reason, I think my job is to reintroduce my people to the real breads of Crete. You see, bread is the key to life. It is one of the secrets to our ancestors' good health, and it has been eaten the same way as far back as the Minoans five thousand years ago. But not the bread you buy packaged in the store. That bread is useless. You have to learn to use these healthy flours to make your own bread."

At Stelios's recommendation, I visited a nearby bakery that uses his flour to make rusks, the authentic Cretan twice-baked bread. Rich, with a nutty aroma and chock full of something recognizable as a *whole* grain, this bread puts other loaves I have known to shame. Unlike most breads, which I would call "filler," this one stood up proudly as "food." Suddenly I could understand why Stelios was a bread zealot. I thought back to the ciabatta that Tanya said was her favorite. Indeed, on the surface, that bread was the same toasty brown as this Cretan rusk. But break it open (or rather *pry*, if it is the Cretan bread), and you will find no resemblance. The ciabatta's center is a mass of fluffy fast-release white sugar as compared to the complex inner world of this rusk.

## What Is a *Whole* Grain?

Until recently, there has been no official definition for *whole grains* and no easy way to tell the imposters from the real thing. In 2006, however, the FDA helped clarify the matter by offering an official definition: Whole-grain foods should contain the three key ingredients of cereal grains—bran (the fiber-filled outer part of the kernel), endosperm (the inner part and usually all that is left in most processed grains), and the germ (the heart of the grain kernel.) *In addition, these three ingredients*

Indeed, most of the home-cooked meals that I enjoyed on the island were accompanied by equally wholesome breads. At first the hard, fibrous consistency seemed like a bit too great a challenge for my jaw. However, I quickly discovered that dipping hunks of the rusk in lemon and olive oil, wine, or some other delicious sauce was a perfect way to soften them and enhance their complex flavor. From a nutritional standpoint, this also made perfect sense as the acid in the wine or lemon serves to break down the phytic acid in the bread, thus improving the absorption of zinc, magnesium, and B vitamins, as well as other nutrients. Perhaps my favorite way to enjoy rusks is in the form of *dakos*, which are essentially the pizza of Crete. However, unlike Tanya's pizzas that consist of a white-flour crust topped with sweetened tomato sauce and loads of cheese, these *dakos* are made from a whole-grain rusk topped with fresh crumbled feta, parsley, chopped tomatoes, and a drizzle of lemon and olive oil.

## Cretan Cheese as a Spice

I always thought of feta as being the archetypal Greek cheese. However, by the end of my stay on Crete, I had been introduced to a vast array of local cheeses with long tongue-twister names such as *anthotyro* or *kefalotyri*. They ranged from exceedingly soft and mellow (almost like yogurt) to hard and pungent. While these cheeses are high in saturated fat, if used sparingly (almost like a spice), they are a perfect complement to other Cretan foods. I ate them as sprinkled toppings on a dish or in paper-thin slices with bread and beans. Several food researchers have hypothesized that the isoflavones in the horta, wine, and olive oil may block the unfavorable effects that the saturated fat may have on the lining of arteries. In addition, since the animals that produce the dairy are grazing on wild greens, their milk offers a decent source of omega-3 fats.

## What Is the Carb–Heart Disease Connection?

At first glance, it might seem surprising that the whole grains in Stelios's shop have very much to do with preventing or treating heart disease. Most of us associate heart disease with fat intake, not with carbohydrates. In reality, the *type* of carbohydrate we eat is as critical a factor in heart disease as it is in diabetes. Slow-release carbohydrates are protective against heart disease, whereas fast-release ones will build up plaque in our arteries.

How can this be? As is the case with all the inner workings of our body, there are probably many factors at play. First of all, most whole or unprocessed carbohydrates, including Stelios's bread, are filled with fiber. This fiber binds to the cholesterol in the gut and prevents its absorption into the bloodstream. Second, unrefined carbohydrates have slow-release properties and, therefore, cause the body to release less insulin, a hormone that in high concentrations seems to damage arteries. In addition, unrefined carbohydrates (unlike refined ones) are not readily converted into very low density lipoprotein (VLDL), triglycerides, and a fat-glucose compound called *triacylglycerols*. These substances have a way of working their way into the lining of arteries where they cause inflammation and arterial wall damage.

Bread made with Stelios's type of flour also has a much greater nutritional value than bread made with refined white flour. The traditional whole-grain loaf is a wonderful source of vitamins, minerals, antioxidants, fiber, protein, and phytochemicals such as lignans, which all play an important role in heart health. Magnesium, for example, is critical for the proper functioning of our heart muscle, while B vitamins, vitamin E, and folate are involved in many of the body's antioxidation reactions. Finally, this whole-grain bread is filling and relatively low in calories, both qualities that help to keep inches off the waistline. (This will ultimately decrease one's risk for prediabetes and metabolic syndrome.) Given all the benefits of eating whole grains, it would seem that Stelios is on the right track. His crusade to put authentic grains back into the rusks should go a long way to protect the arteries of his fellow Cretans.

## Legumes: Another Heart-Healthy Carb

During my time on Crete, I ate small mezes featuring lentils, black-eyed peas, red and white beans, as well as fresh favas. Sometimes these beans were toasted, sometimes added to stews, but most often they came floating in a pool of fresh herbs and olive oil. Legumes have the same cardioprotective effects as whole grains, and they make a terrific trio with the olive oil and wild greens. In analyzing the results of the recent NHANES (National Health and Nutrition Examination Survey), it appears that men and women who eat legumes four times or more per week have a 22 percent lower risk of heart disease compared with those who eat beans less than once a week. Once again, it is important to eat those beans!

## Starches: A Note on Portion Size

When talking about grains and legumes, portions are an important consideration. The portions of hand-rolled pasta that you get on Crete are more on scale with what we in the United States consider to be appetizer size. Even the ceramic plates on Crete are smaller than your typical North American ones. This helps explain our problem with portion control. The bottom line is that while whole grains and legumes have many benefits and are an important part of the Mediterranean diet, we need to be mindful of the amount we eat. In general, for most adults, I recommend that you consider ¾ to 1 cup of a grain or legume or one slice of bread as your portion for a meal.

## Healing Spirits

On my map, Nikki had highlighted the town of Vamos, a restored twelfth-century hamlet located near Hania. There, in the town center, was Taverna Sterna tou Bloumosifi where Nikki assured me I would find truly authentic Cretan food. Indeed, the restaurant's cook was an elderly woman with a sun-creased face and strong hands. The fact that she was wearing the traditional all-black getup and head scarf was a fairly good guarantee that the food would be equally traditional.

By contrast, my waiter Nikos looked like he belonged in a hip sidewalk café circa 1975. His long salt-and-pepper hair was tied back at the nape in a pony tail, and under his waiter's apron, he wore a rock-and-roll T-shirt and bell-bottom jeans. After serving me a perfect meal of freshly baked rusks, grilled sardines, horta, fava puree, and the most delicious meat and vegetable hand pies, Nikos plunked a rather large carafe of raki on the table. I tried to refuse politely but quickly realized that he looked offended. In his opinion, the true lifeblood of Crete was this powerful eau de vie made with grape must, the leftover from wine making. "You should drink this at every meal; it will make your heart strong," he announced as he poured me a second glass. "I am much older than you think," he continued, implying that he had maintained his youthful appearance and his thick ponytail by drinking a large carafe-full each day.

Indeed Nikos was not alone in believing that the spirits of Crete were a factor in preserving heart health. In small tavernas throughout the island, I had seen men in their eighties or nineties enjoying a breakfast of dark bread dipped in red wine and olive oil, and one small glass of wine accompanies most lunches. In addition, nutrition research does seem to support the idea that wine is good for the heart and might also be good for the brain. (The debate still rages among scientists whether it is the grapes, the alcohol, or a combination of the two.) Certainly wine is a good source of polyphenols, which may specifically protect the heart by preventing inflammation and clotting and by helping to relax the arteries.

However, as I stood up from my delicious meal in Vamos, the ancient stone sidewalk began to gently undulate under my feet and my immediate thought was *moderation*. Contrary to Nikos's advice, when it comes to alcohol, more is not better. All the research seems to show that one glass per day is good for the heart, but the benefit quickly turns into a risk the more you exceed this amount.

### Greek Yogurt for Dessert: Continuing the Love Affair

During my visit, I fell in love with the standard Cretan dessert: a simple combination of Greek yogurt drizzled with a little honey or some fruit preserves called "spoon sweets." Upon my return home, I missed this yogurt's creamy consistency until a friend taught me that Greek yogurt can be crudely approximated by simply straining regular yogurt. Ever since, I have been making my own imitation Greek yogurt by putting regular store-bought yogurt in a colander lined with cheesecloth and allowing it to drain for three to six hours.

# Where Have All the Windmills Gone?
## A Lesson in *Siga Siga*

"*Siga siga*," cautioned Manuelis Siganos as he watched me attack the freshly made mezes that were just laid out on my table. Since it was not yet tourist season, I was eating in the only open restaurant in the town of Tzermiado, high on eastern Crete's Lasithi plateau. I had reached the town several hours earlier after driving a vertical distance of 3,000 feet along hairpin turns. The plateau is a large circle of fertile fields surrounded by a chain of tiny towns. Because of its unique geography, this place feels like an island within an island, and daily life here has probably not changed very much over the past several hundred years.

Manuelis was one of the rare English speakers in Tzermiado and seemed to have the monopoly on the town's tourist trade. He had greeted me at my hotel and now reappeared as my waiter in the nearby taverna where he helped me order my mezes off of the all-Greek menu. I was starving, and when the food arrived I enthusiastically attacked the first plate. To my surprise, Manuelis leaned over from the neighboring table and scolded me: "*Siga siga*" ("slow slow" in Greek). Clearly he thought I was eating way too fast. I picked my head up and surveyed my dining mates in the taverna. Immediately I felt embarrassed. In the time that I had wolfed down a small plate of horta, some snails, a plate of fava, and two vegetable pies, the four people at the table next to mine had eaten an eighth of their food. Taking their cue, I slowed down. Instead of shoveling, I relied on large gulps of air and deep breaths to fill the void between bites. As I did this, I noticed that the heavy feeling in my esophagus began to ease, and I started to actually taste each mouthful.

When Manuelis heard about my reason for visiting Crete, he too had a lot to say about why his ancestors had such healthy hearts. Less than fifty years ago, he told me, Lasithi was covered in windmills. There used to be thousands of them. They would pump water from deep in the ground to irrigate all the fields on the plateau. Manuelis pointed to the wall where an old sepia-colored photograph depicted the same landscape I had admired earlier from my hotel room. However, there was one major difference. In the photograph, the patchwork of fields was dotted with the white sails of windmills, whereas the landscape I had seen several hours earlier had none.

Sadly, continued Manuelis, most of these windmills had disappeared. Farmers discovered that diesel engines work much more efficiently for pumping water, and they take less physical energy. Just a flick of a switch is all you need for diesel engines. With windmills, the water flows slowly, you have to struggle to put on the sails, and you are constantly walking the property to maintain them. But Manuelis missed the sight of these graceful, white, bird-like objects. Furthermore, he realized that although their upkeep is time consuming and their output less impressive than a motor, these windmills offer many advantages. They are picturesque, affordable, and nonpolluting and, perhaps most importantly, they help maintain general physical fitness. In fact, he was convinced that these wind machines were at least partially responsible for the longevity and good health experienced by his grandfather's generation. "The young people are fat," he said, "because we are lazy and don't work with the windmills." As a result, Manuelis has started a project to bring the windmills back to the plateau.

Walking back to my hotel, I passed a couple of Manuelis's white-sailed installations and thought about his story. I appreciated how the loss of the windmills was a metaphor for a changing way of life. Whether one is discussing eating patterns or irrigating fields, "siga siga" might be another key to Cretan good health.

## A Study in Synergy

Back in San Francisco, I surveyed my travel notes. Everyone I met on Crete seemed to have their own opinion about which foods were particularly heart healthy. Recalling my conversation with the cardiologist Paul Ridker, all of them seemed right. Almost every food I had sampled could stake a claim to being the key healing ingredient. Certainly the omega-3 fats in the fish, snails, wild greens, and dairy had important anticoagulant effects, while the antioxidants in the fruits, vegetables, olive oil, and wine had strong anti-inflammatory properties. Then there were the insulin-lowering, slow-release carbohydrates that were found in the bulgur, the lentils, and the whole-grain rusks. All three categories of foods served to prevent atherosclerosis, a buildup of dangerous plaque in the arteries.

But which of these foods were most responsible for making Crete a famous heart disease cold spot? Trying to sort this out, I pulled out a research paper written by the Athens-based physician and epidemiologist Antonia Trichopoulou that was published in the *New England Journal of Medicine*. Her study team had distilled Crete's traditional diet into a nine-point Mediterranean diet scale—the more points one has the better. What they found, after surveying over 22,000 Greek adults, was that a mere two-point hike on the scale was associated with a 25 percent reduction in total mortality! I figured that this paper would help me determine which foods on the scale were the most protective.

I studied the article carefully. One of the tables in the paper gives a list of the most common Cretan foods and correlates these foods with their "hazard ratio for death." That is a scary-sounding, yet scientific, way of describing how much a certain food is linked with mortality. A hazard ratio greater than one means that eating that particular food puts you at an increased risk for heart attack or death, a ratio of less than one means it will decrease your risk, and a ratio of exactly one means that your risk is zero—neither increased or decreased.

I looked at the table fully expecting dramatic results. To my surprise, all these supposedly healthy foods such as fish, vegetables, fruits, beans, grains, and olive oil all had a hazard ratio for death that was one or slightly less than one. Furthermore, foods generally considered unhealthy such as saturated animal fats and sweets all had a ratio of only slightly more than one. So how could it be that the Mediterranean diet as a whole could be so protective against heart disease and overall mortality, and yet individual parts of the diet did not stand out as being *extra* protective? I mused over this a bit and wondered if there was something wrong with Trichopoulou's research. Then I suddenly realized that this finding made all the sense in the world. As a matter of fact, it was the perfect illustration of why we needed to focus on indigenous diets rather than on specific nutrients or individual foods. *The true healing effects of the Mediterranean diet cannot be distilled down to a handful of specific ingredients. It is in the food synergy, the ways that the foods interact in the recipes, that the real magic takes place.*

Thinking back to my travels around Crete, I realized that the whole week was a lesson in food synergy. In truth, the basic ingredients were pretty consistent everywhere on the island. They were the fresh foods of that season: fava beans, lentils, wild greens and fresh green herbs, tomatoes, olives, artichokes, fish and octopus, hard barley rusks, goat and sheep cheeses, snails, lamb, chicken, yogurt, figs, lemons, oranges, raisins, honey, and pool after pool of the most delicious olive oil all washed down with a glass of homemade red wine or raki. But then, thanks to the recipes, there was another level of complexity. For example, the olive oil and lemon increased the availability of the nutrients in the greens; the wine and lemon broke down the phytic acid in the rusks; and the antioxidants in the greens and olive oil prevented the lipid peroxidation of the saturated fats in the cheese. And finally, there was the interaction between the half-dozen (or more) mezes that were eaten at each meal—each one contributing its own little world of healing ingredients. Given these layers of combinations, it is no wonder that I could not pinpoint *one* key Mediterranean food; it is the seemingly endless interplay of all these simple foods that offers the real medicine.

## Eating Meze~Style

*Tanya loved the recipes that I brought back from Crete and even mailed some home to her parents. She decided that substituting her favorite foods with more indigenous versions would be easy. The dakos and the bireiki were good replacements for the pizza, and she liked the lentil stew so much that it became a weekly standard. Tanya came back to see me six weeks after starting on her new cooking plan. She had lost five pounds, and her blood pressure had come down several points. She told me that starting an exercise routine had turned out to be challenging, given that she had two small kids. However, she had been walking for at least thirty minutes daily and was doing the treadmill at the gym twice a week.*

*A month later she came back to get the results of her fasting labs. Her LDL had dropped a critical forty-five points, her HDL had come up ten and her C-reactive protein level was in the "low risk" category. Rejoicing, she relayed a funny story that her mother had told her during their last phone conversation.*

Several days after receiving the recipe for lentils and wild greens, her father was seen wandering around a plot of undeveloped land near the diner, carrying a large bag. Stooping occasionally, he would pull up a weed and give it a taste test. Apparently the recipe had triggered some childhood memories of foraging for greens in his native village in Italy. That evening he cooked up his haul with some olive oil, and it tasted surprisingly good. Tanya laughed, telling me that her father was now considering adding "weeds" as a side dish on the diner's menu.

## Three Steps to Eating Like a Cretan

### Step 1 (Basic):

- Eat at least *four* servings per day of in-season vegetables. (A serving is 1 cup raw or ½ cup cooked.)
- Think olive oil. Use extra virgin olive oil as your main added fat (be mindful of the calories).
- Replace white pastas, white or whole-wheat bread, rice, and starchy potatoes with more slow carbohydrates. These include pulses of all varieties (chickpeas, lentils, broad beans, white beans, fava, etc.), bulgur, whole-grain cracked wheat, whole-grain bread, waxy (new) potatoes, yams, and whole-grain pasta.
- Enhance your synergy. Eat red meat in stews with grains and vegetables.
- Limit the red meat. In general, have no more than one red meat dish per week.
- Treat cheeses like a spice or a condiment.
- Watch that portion size. Use appetizer-size plates to control portion size; try to limit your starches to ¾–1 cup per meal.
- Do some form of aerobic exercise for at least thirty minutes, four days a week. Fast walking is an excellent choice.
- Cook at least one recipe per week from the Crete recipe section.

**Add Step 2 (Intermediate):**

- Go wild. Have at least three servings a week of wild-type greens, including dandelion, Swiss chard, kale, sorrel, carrot greens, and arugula.
- Eat at least *five* servings per day of in-season vegetables. (A serving is 1 cup raw or ½ cup cooked.)
- What your meat eats matters. Choose free-range, grass-fed (not grain-fed) beef, lamb, chicken, and other meats. Eat free-range eggs.
- Snack substitutions. Replace all processed snacks with a piece of fruit, some cut-up vegetables, or a small handful of nuts and seeds.
- More food synergy. Enhance your meals with fresh squeezed lemon and organic yogurt.
- Get an omega-3 charge. Eat three ounces of fish at least once per week.
- Herbal pharmacy. Add Mediterranean herbs to your meals for flavor and heart health. It's easy to add chopped oregano, parsley, chives, garlic rosemary, and more to salads, sandwiches, and main course dishes.
- Spirits in moderation. Drink no more than 1 glass of wine per day. (Grape juice or pomegranate juice might work just as well.)
- Choose simple desserts that are fruit-based (for example, figs drizzled with honey and walnuts.)
- When eating out Mediterranean-style, verify that your meal is cooked with olive oil. Order appetizer portions of pizza and pasta, and make sure that your main dish has a high proportion of vegetables.
- Do some form of aerobic exercise for at least thirty minutes at least five days a week. Fast walking is an excellent choice.
- Cook at least two recipes per week from the Crete recipe section.

**Add Step 3 (Advanced):**
- Eat at least *six* servings per day of in-season vegetables. (A serving is 1 cup raw or ½ cup cooked.)
- Share your table. The long-living Cretans had strong family connections and ate meals together. Don't underestimate the benefits of community to your health.
- Make some mezes. Try to sample a variety of simple dishes at one meal. This enhances your food synergy and delights the taste buds.
- *Siga siga.* Chew your food completely and eat slowly. This often results in eating less overall.
- Modified fasts for your health. Consider observing occasional week- or month-long modified fasts as these have been shown to be beneficial for your health. During this time give up all rich foods, including sweets, baked goods, dairy products, and red meat.
- Be bold. Try to make your own Cretan rusks.
- Exercise daily for thirty minutes. In pre-1970s Crete, most people got around by foot!
- Cook at least three recipes per week from the Crete recipe section.

# 7. Iceland: A Cold Spot for Depression

"Why Iceland?" asked my friends on hearing my travel plans for mid-July. After all, in the middle of a typical freezing and foggy San Francisco summer, the last place I should want to go is a country famous for its glaciers, a place where the midsummer temperature rarely reaches 65 degrees. The truth is if it were not for my patient Christine and my search for a depression cold spot, I may never have chosen Iceland as a vacation destination. But then again, it is probably this line of thinking that has helped an island nation with barely 300,000 citizens maintain its cold spot status.

In this chapter we focus on depression, a disease high on the list of modern chronic health problems. After heart disease, depression is the single greatest cause of disability worldwide. In the United States, the lifetime risk of experiencing depression is over 20 percent in women and 12 percent in men. Like the other health problems discussed in this book, depression does not occur uniformly in all populations. As a matter of fact, there are huge variations from country to country and city to city, with some areas having as much as six times the depression rates of others. Researchers have tried to identify factors that can explain this variation. While they can all agree that there are a multitude of issues at play, most of them feel that diet is an important factor.

# Do I Have Depression?

Depression is not just a passing case of "the blues." It is characterized by a deep sense of sadness and hopelessness that lasts months to years with symptoms ranging from mild to severe. A diagnosis of major depression requires that symptoms are significant enough to interfere with a person's relationships and daily activities. There are a whole range of factors that can trigger a depression: some, like the death of a loved one, are external triggers, while others are more internal, caused primarily by an imbalance in brain chemistry. Similarly there are many possible treatments for depression, including (but certainly not limited to) medication, psychotherapy, and lifestyle changes such as diet and exercise.

Signs of depression include the following:

- Loss of interest in activities that were once engaging or enjoyable
- Loss of appetite or overeating
- Loss of emotional expression
- A sad, anxious, or empty mood
- Feelings of hopelessness, worthlessness, or helplessness
- Social withdrawal
- Low energy level
- Sleep disturbances, including insomnia or oversleeping
- Trouble concentrating
- Alcohol or drug abuse

At first glance, Iceland may seem like an unlikely depression cold spot, especially if one considers that Icelanders spend a good part of their freezing winter months in continual darkness. However, the fact that a number of studies highlight this island as a land of distinctly undepressed people prompts me to set off and explore this circumpolar nation. As I prepare for this journey, a diverse selection of researchers help me understand the intricate connections between mood and food. They include an MIT endocrinologist, an NIH epidemiologist, and a University of Wales psychologist. Once I arrive on the island, my conversations with an Icelandic physiologist from the University of Reykjavik, a fishermen, a hot-house vegetable farmer, a bed-and-breakfast owner, and a supermarket clerk (as well as my own gastronomical experiences) help me understand how indigenous Icelandic foods come together to make an antidepressant diet.

## Meet Christine

*Christine's first appointment in my office was on a rainy morning in mid-January. She was a tall, serious-looking woman dressed in a business suit and carrying a large leather briefcase. Within seconds of our meeting, she snapped open the case and presented me with a lengthy medical file. Her manner gave me the sense that she was used to efficiently run business meetings and expected results. However, once she began to tell me her story, Christine quickly lost her composure. Dabbing away tears with a tissue, she told me that she had been struggling with depression since she was a teenager. She had made an appointment with me because her latest antidepressant medication had stopped working and she wanted me to prescribe a newer one. Lately, Christine had been feeling a constant sadness and edginess: she did not want to go out in public, and she found herself snapping at coworkers, her husband, and her young daughter. To make matters worse, she told me that she was once again binging on cookies and croissants. Instinctively, she felt that these sweet and starchy foods might soothe her or make her happy. In fact, they made her feel better for a brief period, and then suddenly, she would feel even more depressed than she had before she had started eating. At this point she was worried that the weight gain caused by her carb cravings might be triggering other health problems since a couple of recent blood pressure readings had been on the high side.*

*Christine stopped me midsentence when I began discussing how the right diet plus exercise can help to treat depression. She had been told in the past that her depression, her weight, and her blood pressures might be improved with lifestyle changes, but she felt that this was too hard. After all, she had tried many diets,* and they had all failed; she also disliked exercising or breaking a sweat. Food was one of her few comforts, so she preferred to just take medication even though most antidepressants only seemed to help for a couple of months before losing their effect. By the end of the visit, I did what many of her doctors had done before: I prescribed Christine one of the newest antidepressants on the market and asked her to come back soon and to call if the medication did not help.

Four weeks later Christine returned. She reported that the latest medicine had worked briefly, but its effect seemed to be wearing off. Once again, she was starting to spiral downwards and feel desperate. Unsure what to do next, I referred her to an excellent psychiatrist and asked her to make an appointment with him as soon as possible. My colleague did feel that there were other antidepressants that Christine could try; however, he had little hope that any of them would work effectively unless she changed her diet and increased her physical activity. He explained to her that the benefits of even the most powerful antidepressants are quickly overshadowed by an unhealthy lifestyle, and he encouraged her to explore how she could alter her diet to treat her depression.

Christine felt backed into a corner. Reluctantly she called me and told me that she was ready to discuss her diet; however, she cautioned me that she would only be willing to change certain things. After all, she was raised in a family of "dough dogs" and meat eaters, where bread and meat were a prominent part of every meal. Some of her happiest childhood memories included family forays to the local Dunkin' Donuts or weekend dinners at the little A&W stand. In addition to being a "dough dog," Christine was a self-proclaimed "vegetable hater." She rarely ate them unless they were disguised in butter or deep-fried batter. Salads, to her, tasted like a "bowl of grass." Given her love of bread and meat and her dislike of all things green, Christine challenged me to find foods that would help her fight her depression.

## Food and Mood: What Is the Connection?

Most of us intuitively know that there is a connection between what we eat and how we feel. If not, why would we use such expressions as sugar high, sugar low, food coma, or caffeine buzz? We instinctively know that missing meals usually makes us feel grumpy; overindulging in something can make us feel dull or guilty, whereas eating something we perceive as healthy leaves us with a sense that all is right with the world. But how does our food affect these emotions? Food chemists, psychologists, and other scientists have devoted their careers to trying to understand these links. Needless to say, it is a complex topic, and many of the connections between food and mood remain a mystery. Nonetheless, as I surveyed the literature and spoke with experts, it appeared that there are four main ways that food can help trigger or treat a depression. I will categorize these four food-mood connections as sensory, symbolic, diversionary, and neurochemical.

First of all, the *sensory* experience of food (taste, smell, and texture) activates neurological pathways in our brain that affect mood. It is fascinating to think that the experience of sweetness, for example, can have a physiologic effect that is similar to a narcotic pain medication. Next, there is the *symbolic* value of the food. Some foods make us happy because we associate them with a positive experience, be it a special holiday, a wonderful vacation, or a family tradition. Conversely, certain foods can make us feel sad or guilty because we associate them with negative factors. Christine mentioned to me that she breaks into a wide grin whenever she drives past a doughnut shop or an A&W stand—these foods have a positive association, and therefore, they help (at least temporarily) to enhance her mood. Food can also serve a *diversionary* function. In other words, for some people, the act of eating can be a distracting activity, especially when they are stressed. It is not uncommon to have a troublesome thought and suddenly find oneself in front of the open fridge. Unfortunately, while eating might briefly distract one from stressful thoughts, most studies show that this benefit is short-lived. Overall, food is not an effective way to relieve stress. Finally, as we shall see in numerous examples throughout this chapter, nutrients in certain foods can have direct *neurochemical* effects on the brain. Like psychiatric drugs, these nutrients act as hormones, enzymes, and other neurotransmitters that send messages throughout our nervous system. In this instance food is literally a drug.

## Sweets and the Sensory Connection Between Food and Mood

Scientists have used sophisticated brain-imaging techniques called PET scans to show that sweet tastes and narcotics activate the same pain receptors in our brain. This explains why infants undergoing painful procedures in intensive care units seem to get as much pain relief from a sugary solution as they do from narcotic medication. Interestingly, it appears that pain is blocked by the *sensation* of sweetness, rather than any molecular activity of the sucrose, since noncaloric sweeteners such as aspartame seem to have the same effect.

## Chocolate and the Symbolic Connection Between Food and Mood

In one study researchers showed that people who perceived chocolate as being "bad" for them were more likely to feel depressed after eating it; whereas, those who thought of it as "healthy" were more likely to feel positive after having a piece.

Given the different ways that food and mood interact, it was a daunting task to give Christine specific dietary suggestions to treat her depression. Clearly, I needed to help her find a diet rich in nutrients that would have a positive effect on mood. In addition, the foods needed to conform to Christine's taste buds (sensory) and trigger pleasant associations (symbolic). What I needed to find was a cold spot for depression with a culture and culinary tradition that would meet Christine's approval.

Comparing rates of depression around the world, the small island nation of Iceland consistently emerges as a depression cold spot. One recent European study looked at depression rates of people over age sixty-five and concluded that those who lived in Munich, London, Amsterdam, Verona, and Berlin had the highest rates, whereas Icelanders had the absolute lowest. Other research has supported these findings. Consistently, Icelanders seem to experience less depression in all of its forms, including seasonal affective disorder (SAD, or "winter blues"), post-partum depression, and bipolar disorder. Icelanders also fare well in other quality of life indicators. When compared to North Americans, they have almost half the death rate from heart disease and diabetes, significantly less obesity, and a greater life expectancy. In fact, the average life span for Icelanders is among the longest in the world.

Needless to say, I was initially surprised to learn that Iceland was such a promising cold spot for depression. Was this not the land of permafrost, cataclysmic volcanoes, endless darkness in winter, and the unsettling midnight sun in summer? Was this not the land of sagas—tales full of murder, deceit, and mayhem? How could Icelanders help but be depressed? After all, Finland, a relatively close neighbor to Iceland with a similar climate has notoriously high rates of depression and suicide. Furthermore, other countries at much lower latitudes fare much worse when it comes to their depression statistics.

I am not the first person to be puzzled by these findings. There are a number of articles published in medical journals and popular magazines that speculate as to why Icelanders are generally so positive. Some propose that there is a cultural bias to the survey questions and that an Icelander would never admit that he was depressed. As a matter of fact, the day I was flying out of Reykjavik, the popular weekly English language magazine *The Grapevine* had a series of articles and editorials discussing Iceland's status as a happy nation. The front cover of the paper had a caption reading, "Happiest Nation on Earth," with a photograph of a man using his fingers to pry his unwilling mouth into a grin. The feature article was written in slighty stilted English, but its message was clear enough: "It does not matter if you are an Icelander who just so happens to be looking for some rope to hang himself with—when asked how he is doing, he's going to respond with the classic Icelandic, 'Really good, thanks.'" While bias is a possibility, most of the researchers whom I talked with assured me that the questionnaires are specifically designed to overcome cultural differences. Furthermore, keeping a tight upper lip is a quality shared by many Northern European countries, so this fact alone could not explain why Icelanders scored so much lower on the depression scales than other Northern Europeans.

Others believe that the low rate of depression (as well as the great longevity) is in the genes. Iceland was first inhabited by Vikings over 1,000 years ago, and until very recently, few others have immigrated to its shores. As a result, the island has maintained a closed gene pool—so closed, in fact, that scientists have kept excellent genealogical records of the entire population. Given this situation, perhaps what we are looking at is just a case of survival of the fittest and least depressed. In other words, those Vikings who were less prone to the blues were more likely to have survived until child-bearing age. If depression is at least partially a hereditary trait, this natural selection would have eventually produced a nation of depression-free individuals.

This gene theory made perfect sense to me . . . until I came upon the work of Jóhann Axelsson, PhD. A prominent physiologist and depression researcher at the University of Reykjavik, he is one of the country's foremost experts on food and depression. After reading Axelsson's articles, I realized that the *real* explanation for his country's low depression rates might lie somewhere other than genetics. In fact, it was starting to look like diet, once again, was playing an important role. Intrigued, I decided that it was worth a trip to Iceland to find out more.

## The Viking Scientist

Jóhann Axelsson embodies the word *cheery*. I would even go so far as to say that he is the poster child for depression-free Iceland. On my first day in Iceland, he and his wife, Ingur, invited me to join them at their favorite restaurant so that I could sample some high-quality Icelandic fare. This was my first time meeting Dr. Axelsson, but after a bear hug and several minutes of his witty jokes, I felt like we were old friends. Despite the late hour, it was still as bright as high noon, so they proposed a predinner tour of the city. While Ingur negotiated the tight back streets of Reykjavik in their tiny car, Jóhann pointed out his favorite landmarks and gave me some tidbits of history about the city and about himself. He had grown up in the north of Iceland in a small port town and had come from a long line of fishermen. Breaking from the traditional family business, he had decided to go to Paris to study art. "Somehow, instead," he said, "I ended up studying the brain." He had many opportunities to build a science career in Europe, but missing his homeland, he eventually decided to come back to Iceland to teach and set up a depression research lab. As he spoke, it struck me that although he had not worked on the sea since boyhood, he still had the ruddy cheeks and thick gray beard of a Viking mariner. In fact, if you were to replace his tailored blue suit and black lace-up shoes with fur and hides, he would be the spitting image of the Viking wax models that I had visited earlier in the Reykjavik Saga Museum.

When we finally arrived at the restaurant, it was almost 10:00 p.m. and the sun was finally developing that soft rosiness that I usually associate with the early evening. Perusing the menu, which was written in Icelandic, I randomly selected something called *blóÕmör*. I had no idea what it was, but I liked all the accents and the capital O midword. In unison Ingur and Jóhann cautioned me that this one might not be so appetizing to my American palate (I later learned it was blood sausage) and steered me toward the salted cod, turnips, and potatoes. While we waited for our food, I asked Axelsson to tell me about his research.

## The Saga of Two Icelands: Testing the Gene Theory in the Perfect Depression Lab

It was the "SAD paradox" that first drew Jóhann Axelsson to his field of study. He wondered how it was possible that Iceland could be so close to the Arctic Circle and yet have such low rates of seasonal affective disorder and depression. Initially, he too believed that the explanation must be genetic and set out to prove this through a series of experiments. However, like so many Icelandic stories that are filled with twists and turns and unexpected endings, this too turned out to be quite a saga.

Jóhann told me that between 1870 and 1914, Iceland was plagued by enormous volcanic eruptions that covered much of the cultivated southeast region of the country. Traumatized and facing starvation, around one-fifth of the population emigrated to Canada and the United States. Many of them settled on the shores of Lake Winnepeg in Canada's Manitoba province, an area now known as "New Iceland." There they lived as a close-knit group, preserving the customs, the language, and the cuisine of their homeland. To this day, Canadian settlements continue to thrive, and a significant proportion of the population can trace their ancestry directly back to the original Icelandic settlers who moved to that area.

Axelsson realized that this mass emigration and the existence of "New Iceland" provided him with a rare research opportunity. Here were two genetically similar groups, living in different places at roughly the same latitude. If he compared the rates of depression and seasonal affective disorder in these groups and found them to be equally low, then he could finally prove that genes were largely responsible for Iceland's low depression rates.

To his surprise, however, this is not what he found. After interviewing both groups, he learned that the Canadian Icelanders had depression rates that were similar to other Canadians—almost twice as high as the rates seen among native Icelanders. They also had a shorter life expectancy and suffered more from heart disease and obesity. Since both Icelandic groups shared a similar genetic makeup and a similar latitude on the globe, there must be something else that could explain why one group experienced depression more frequently than the other. What could it be? When I asked Axelsson this question, he winked at me and held up a piece of salted cod on his fork. "I can't be certain," he said, "but all evidence points toward the food."

## Beyond Putrified Shark

When I started to investigate the actual foods that set Icelanders apart from more depressed nations, I must admit that I became nervous. Many of this country's indigenous foods were not likely to have a broad appeal. These included the *hákarl* (putrefied shark that is buried for months before being served), the *svið* (halved singed sheep's head), the *Hrútspungar* (pressed and pickled lamb testicles), and of course, the *blóÕmör* (blood pudding). As one Icelandic American friend gently put it, "I do not think that a book full of recipes for whale meat and fermented shark is going to do so well. . . ." Fortunately, as I ate my way around the island, I realized that many of the key ingredients and recipes found in the Icelandic indigenous diet were actually quite acceptable to our North American palates.

## A Kettle of *Fiskur*

First and foremost, there was the fish, or *fiskur* as it is called in Iceland. At 225 pounds of fish per person per year, Icelanders top the charts when it comes to per capita fish consumption. (Japanese come in a distant second at 147 pounds per person per year, while in the United States, we eat a lowly average of 48 pounds.) Ask an Icelander how often he eats fish in a week, and you get a blank stare. You might as well ask him how often he breathes. The answer is, "Too often to count." Indeed, outside of Reykjavik, it seems that everyone is involved somehow in the fishing industry: fishermen, boat mechanics, fish canners, fish farmers, and so on. Paintings of fish adorn the walls of most restaurants, and a string of dangling dried fish is considered a perfectly appropriate doorway ornament.

Fish are also a favorite topic of conversation. In a small café near Lake Myvatn, I met a fish farmer named Robert Boulter and his wife, Inga Karlsdóttir. They were on their way to visit Inga's parents who still run the family dairy farm on the north coast. As their two children fought over who would get the larger slice of whole-grain potbread layered with smoked salmon, Robert and Inga chatted with me about fish. Robert, for one, was amused by the fact that the rest of the world has suddenly jumped on the omega-3 bandwagon. To Icelanders, the effect of fish on health and mood has long been obvious. In addition to frequent meals from the sea, his mother had always given him a daily teaspoon of cod liver oil to keep him healthy and happy during the long dark winter. Even now, as an adult, Robert notices that he gets a "little bit grumpy" after several days without fish.

## Back to the Saga

As we moved on to a dessert of fresh bilberries and *skyr* (similar to blueberries and cream), Jóhann Axelsson continued his saga of the two Icelands. Once they arrived in the Interlake region in Canada, many of the Icelandic settlers continued to do what they did best—fish. However, there were some factors that made the freshwater fishing in Canada much more challenging than the traditional open-ocean fishing in Iceland. First of all, in Canada, the local fish were small and slim, and therefore, they could easily escape through the holes in the Icelandic fishing nets. Second, while oceans never freeze over, lakes do. Even with the proper equipment and skills, it is a daunting task to chop a hole in the ice, and as a result, winter fishing in Canada yielded a smaller daily catch than it did in Iceland. For both these reasons, New Icelanders have always eaten much less fish—especially in winter—than their Icelandic relatives.

Axelsson noted one final difference between fishing in Iceland and New Iceland. This was a difference that was probably not obvious to the original settlers but was certainly obvious to his research team 100 years later: the whitefish and walleye pike being pulled out of the Canadian lakes were much less rich in omega-3 fats than the Icelandic char and salmon. Given that New Icelanders past and present eat less fish (and less omega-3–rich fish) than their Icelandic relatives, it follows that their *total* omega-3 intake would also be lower. Indeed, when Axelsson compared the blood omega-3 levels of his research subjects in both locales, he found that native Icelanders had levels that were three times higher than Canadians of Icelandic decent. Axelsson believes that these high blood levels of omega-3 fats may at least partially explain the lower depression rates.

## The Omega-3-Mood Connection: A Look at the Evidence

What is it about omega-3 fats that wards off depression? While I have discussed the benefits of omega-3 fats in maintaining a healthy heart, even more research is available regarding their effect on the brain and the nervous system. The dry weight of the human brain is approximately 80 percent fat (the highest of any organ), and a high proportion of these fats are the long-chain omega-3 essential fatty acids known as eicosapentaenoic acid (EPA) and docosahexaenoic acid (DHA). While it is a well-established fact that these fats play a vital role in the developing brains and nervous systems of infants and small children, only in recent years has it become clear that omega-3 fats are also critical for the normal functioning of adult brains. Scientists around the globe are discovering that an omega-3–rich diet is linked to less depression and to lower rates of other psychiatric problems, including bipolar disease, schizophrenia, and aggressive or antisocial behaviors.

### Omega-3 Fats Versus Prozac

Some psychiatrists and other physicians are beginning to prescribe omega-3 fats as a substitute or an enhancer to standard antidepressant medications. So far the majority of clinical trials have shown that at least 1 gram per day of EPA or EPA plus DHA (both are types of omega-3 fats found in fish oil) work better than a placebo in treating patients with depression. However, these fats are probably most effective as *complements* to standard antidepressants rather than as stand-alone medications.

Since fish is one of the best food sources of EPA and DHA, many of the studies have looked specifically at fish consumption and its relationship to depression. In Finland, for example, researchers have compared Finns who are not depressed with those who are in order to learn about differences in their diets. What they found was that it all boiled down to fish. Nondepressed people ate fish more often than their depressed counterparts. The studies concluded that eating fish two or more times a week (even freshwater fish) was linked to a 50 percent lower rate of depression!

In other parts of the world, the fish and mood relationship has also been established. Joseph Hibbeln, MD, a researcher at Bethesda's National Institute of Health has done a cross-national study comparing depression rates with fish consumption. Given what we have just learned, it should not surprise us that the countries with the lowest per capita fish consumption (i.e. West Germany, New Zealand, and the United States) have some of the highest rates of depression and vice versa. Hibbeln then compared fish intake to rates of homicide and post-partum depression and found they were also inversely related. When I asked him why Icelanders scored so low on the depression scales, he answered without hesitation: "It's the fish."

## A Word About Toxins in Fish

Recently, the Harvard Center for Risk Analysis convened an expert panel to determine the risk/benefit trade-offs of eating fish. What they found was that, while it is advisable to avoid the species of fish that contain the highest concentration of toxins, the upside to having at least *two* servings of fish per week far outweigh the downside. This recommendation was made for children and all adults including pregnant and nursing women. My advice is to have the two weekly servings of the "safe" fish on the list and, at the same time, get at the root of the problem by supporting the passage of stricter environmental laws. Our ultimate goal should be to reduce the industrial emissions of toxic substances such as methylmercury and polychlorinated biphenyls (PCBs) that are poisoning the world's waterways.

## Which Fish Are Safe? *

| **Green Light Fish**<br>*Low Mercury*<br>*(Safe to eat*<br>*2–3 times per week*<br>*or 12 ounces per week)* | **Yellow Light Fish**<br>*Moderate Mercury*<br>*(Enjoy a fish from*<br>*this list up to 1*<br>*time per month)* | **Red Light Fish**<br>*High Mercury*<br>*(Do not eat)* |
| --- | --- | --- |
| Abalone (farmed) | Alaskan Halibut | Atlantic Farmed Salmon or Great Lakes Salmon |
| Arctic Char | Blue Mussels | Shark |
| Anchovies | Blue Crab (Gulf Coast) | Swordfish |
| Catfish (U.S. farmed) | Black Cod | King Mackerel |
| Caviar (U.S. or French farmed) | Cod, Pacific | Tilefish (Golden Snapper or Golden Bass) |
| Clams (farmed) | Dungeness Crab | Tuna [Steaks and Albacore (white)] |
| Crawfish | Eastern Oysters | Atlantic Halibut |
| Herring | Mahi Mahi | |
| Hoki | Pollack | Oysters (Gulf Coast) |
| Rainbow Trout (farmed) | Tuna (Canned Light) | Pike |

| Green Light Fish | Yellow Light Fish | Red Light Fish |
| --- | --- | --- |
| Salmon (Wild Alaskan and Californian, canned or fresh) | | Sea Bass |
| Sardines | | Cod, Atlantic |
| Squid (Pacific) | | Marlin |
| Striped Bass (farmed) | | |
| Sturgeon (farmed) | | |
| Pacific Flounder | | |
| King Crab | | |
| Scallops | | |
| Pacific Sole | | |
| Tilapia | | |
| Sand Dabs | | |
| Shrimp | | |

*Information for this chart is compiled from recommendations from the Environmental Working Group and The Green Guide Institute; both are independent, not-for-profit research institutes that provide unbiased information for consumers.

## And Still More Omega-3s . . .

Initially Jóhann Axellson gave fish the full credit for Iceland's low depression rates. But there was a small detail that troubled him. Many of the Icelanders whose blood omega-3 levels he had measured were from the region surrounding the town of Egilsstaðir. If you look at Egilsstaðir on the map, you can see that it is over twenty kilometers from the coast, in the heart of one of Iceland's agricultural zones. When I drove along Route 1 in this area, all I saw for miles was field after field covered with puffy white clumps of grazing sheep. This is mutton country. In fact, for every meal in this region that features fish, there are at least four meals with lamb as their main course. Why then, wondered Axelsson, were the blood omega-3 levels of the Egilsstaðirians as high as those of the coastal Icelanders?

The mark of a good scientist is versatility and a willingness to follow fresh clues, even when these clues suddenly throw you onto an unexpected track. Axelsson and his colleague Gudrún Skúladóttir certainly fit this bill. One spring, at the height of the lambing season, they traveled to Egilsstaðir and set about measuring the omega-3 content of the lambs in the region. To their surprise, they discovered that the muscle and fat from these lambs were very rich in omega-3. In fact, the lambs were almost as good a source of this nutrient as the fish. At first the researchers simply assumed that the lambs (or their mothers) were being fed fishmeal, but they soon realized that even those sheep that were not eating fishmeal had high levels of omega-3 in their tissues. What could possibly account for this?

While hiking through the grazing fields near Egilsstaðir, I got my answer. Everywhere I looked were acres of intensely green, low-lying moss and clover. Kneeling to pick some tender young leaves, a massive clump came out in my hands, and I noticed that there were hardly any roots. These plants have a short life-span due to the thick snow that covers the hills from October to April. However, the fact that these plants are so short-lived makes them an incredibly rich source of another omega-3 fat, alpha linolenic acid (ALA). The Icelandic sheep, whose ancestors were first brought over from Norway in the ninth century, have evolved to thrive on this low-lying grass and to convert the ALA efficiently in their bellies to the long chain omega-3 fats EPA and DHA. So this explained why all the lambs were born loaded with omega-3s: they were nourished by ewes who had eaten island moss throughout their entire pregnancy.

## Offal: The Good and the Awful

While all parts of the Icelandic lambs contain omega-3 fats, the internal organs (*offal*), including the kidneys, liver, intestines, gonads, and brain are the richest source of these nutrients. They also harbor high concentrations of other vitamins and minerals, including iron, folic acid, vitamins A and D, and zinc. Icelanders probably get their biggest boost of omega-3s from lamb offal during the slaughter season that takes place each fall, long after most fair-weather tourists like myself are safely back in their own countries. While I love a good lamb chop, I would be lying if I said that I was sad to have missed this event. Not that I hate organ meats. In fact, there are some traditional recipes using kidney, tongue, or liver that I find quite appealing. However, I was not sure if I could stomach one of the most celebrated indigenous specialties: *svið*, or lamb brain, singed in the "shell" (skull). To my amazement, even small children that I spoke with seemed to regard this as an epicurean delight. (It should be noted that while organ meats are nutrient rich, they are also calorie rich; therefore, I recommend keeping your servings on the smaller side (two to three ounces).

## A Nation Bathing in Omega~3s...

Wild game is yet another source of omega-3 fats in the Icelandic diet, and like free-range domestic animals, they have relatively low concentrations of less healthy saturated animal fats. The most popular game on the island are two species of sea birds: puffins and guillemots. I tried guillemot, a red-meat bird, grilled and served with dark bread. It was memorably delicious. I also had several opportunities to sample puffin breast, but after spending an afternoon watching these fluffy creatures frolicking off-shore in northern Iceland, I couldn't bring myself to eat one. Nonetheless, I have been told by wild game aficionados that they are also quite delectable.

As if the fish, lamb, and wild game were not enough, I soon learned of yet another way that the Icelanders were getting their omega-3s. In a café near Egilsstaðir, I watched two men as they sat together over lunch, sipping from tall mugs. Instead of the standard beer, however, they were both drinking frothy fresh milk. This milk as well as other dairy products (including local varieties of cheeses and *skyr*—a delicious Icelandic version of yogurt) are all unusually rich in omega-3s. When researchers compared whole milk from cows of five Nordic countries, they found that Icelandic cows had by far the highest omega-3-to-omega-6 ratio and the lowest amount of saturated fat. Other studies looking at cheese have had similar results. Apparently that ALA from the clover and moss is also making its way into the milk. I later learned that Swiss Alpine cheese (from Switzerland), which we can buy in the United States, comes from cattle who have grazed on a similar clover and is also is a good source of omega-3s—it has five times the amount of omega-3 fats and 20 to 30 percent less saturated fat than a standard U.S. cheddar.

So while fish was an excellent source of omega-3 fats in the traditional Icelandic diet, it was far from the only one. Between the lamb, the wild game, the cheese, the *skyr*, the milk, and (of course) the fish, the Egilsstaðirians in Jóhann Axelsson's study were practically bathing in omega-3s! In fact, along with geothermal energy, Iceland would do well to claim these fats as one of its greatest natural resources.

## A Plug for Organic Dairy

In addition to having fewer pesticides and hormones, it appears that milk that is labeled "organic" also offers a better fat profile than conventional milk. This might be due to the fact that these cows spend more time grazing in an open pasture rather than being fed corn and other processed feeds. A recent study at the University of Aberdeen compared the omega-3 content in milk from thirty-six organic and conventional dairy farms around the United Kingdom. They found that organic whole milk contained 70 percent more omega-3s than nonorganic whole milk, and the ratio of omega-6 (inflammatory fatty acid) to omega-3 (anti-inflammatory) was much lower in the organic milk.

## Antioxidants for Vegetable Haters

Eating my way around northern Iceland, I was struck by how few green foods were served at every meal. In my usual diet, I am used to having stewed greens, broccoli, or a big green salad with most dinners, and I was beginning to wonder why these healthy Icelanders did not eat more of these foods. When I asked a couple of locals about the notable lack of green veggies in their diet, they scoffed. "Oh those, they are not real food. They are just extras." According to the writings of Icelandic chef Nanna Rögnvaldardóttir, many Icelanders share this sentiment. I was puzzled by this attitude and decided to make a stop at Hveravellir Farms to learn more.

As I approached the farm, I thought that the string of low-lying buildings, each belching out a plume of white smoke, was a small factory or a garbage incineration plant. Coming closer, I realized that what I was looking at was a dozen geothermally heated greenhouses. I wandered into the farm's warehouse where workers were busy crating tomatoes, peppers, and cucumbers. When I explained that I was interested in Icelandic food, the farm's supervisor kindly offered to give me a tour of the premises. Because of the island's short growing season, most vegetables need to be grown in greenhouses. We looked into one of the greenhouses, and I noticed that it was packed with super-sized plants. To maximize production, the vines were trained to grow from floor to ceiling on long metal wires. As we toured, my guide informed me that this small operation produced the vegetables for one-eighth of the entire country. How mind-boggling that this relatively tiny farm was one of the only sources of vegetables for almost 40,000 people!

At the end of the tour, the greenhouse supervisor grabbed a tomato out of a crate and offered me a slice. To my dismay, it was pale pink, hard, and entirely flavorless. Certainly not the ruby red, juicy, sweet organic fruit that I am lucky enough to get in late July in California. Immediately I understood why many of the Icelanders I spoke with were not very fond of fruits and vegetables and did not regard them as "real food." No doubt I would share this sentiment if this was my only produce.

This experience got me thinking: the omega-3s in the Icelandic diet might be having a strong antidepressant effect, but how were Icelanders able to do so well despite eating so few vegetables? After all, most of the food and mood research points to a variety of antioxidants, vitamins, and minerals (which are typically found in fruits and vegetables) as being critically important for maintaining a healthy nervous system. These nutrients help with synthesis of antidepressant neurotransmitters such as serotonin, and they help to protect and regenerate nerve tissue. At this point a large body of research shows that people who are deficient in nutrients such as vitamin C, B vitamins, and folate are much more likely to be depressed.

Once again, I needed to look beyond the obvious. Like the omega-3s, maybe Icelanders were getting their antioxidants, vitamins, and minerals from some unpredictable sources. The first clue came to me while hiking along a mountain path on the eastern side of the country. Suddenly my eye was drawn to the small piles of sheep dung that lined the trail. They were a phosphorescent purple! What could possibly be the source of this color? On closer scrutiny, I realized that tucked amidst the bright green ground cover were thousands of purple berries. They were everywhere. Fearful of poisoning myself, I picked a couple and brought them back to Astrid, the hostess at my bed-and-breakfast, to get an ID. She laughed, telling me that if I waited a couple of hours, I would get a load of them in my dinner soup. They were bilberries, equivalent to our blueberry and similarly full of antioxidants. For the two months when they are in season, Icelanders pick them by the bucketful and eat them fresh in soups, with meat, over *skyr*, and baked into desserts. They then freeze or dry the rest to save for the winter months.

The following day, I asked for bilberries at the local Samkaup supermarket. The woman behind the counter looked at me confused. "Why should we sell them," she asked, "when you can go out and pick them for free?" How impressive. Not only were Icelanders eating these little antioxidant bombs, but they were also getting their exercise by crawling along the mountain paths to collect them by the bucket-full.

The more I chatted about food with Icelanders, the more I learned about other unexpected antioxidants that were making their way into the Icelandic diet. Astrid was one of first people to tell me about the moss.

For centuries, Icelanders have been mixing their native moss with rye or barley flour to bake bread. Historically, this practice was probably a safeguard against the frequent famines that afflicted the islanders. However, in modern times it continues to be accepted as a pleasant flavoring and a healthy additive to a meal. According to Astrid, the kids won't touch it, but the older folks love it and still eat it. I am willing to sample anything edible at least once, and so the next day, while I was hiking, I decided to nibble on some moss. While I cannot say it was fabulous, it did have a vaguely sweet taste, and unlike other greens in Iceland, it was plentiful. Later, in a health-food store in Reykjavik, I saw this same type of moss being sold as "mountain grass" with a little label saying "A natural source of antioxidants!" Indeed, this was a surprising substitute for crunchy vegetables.

## Tea: An Uplifting Drink

Black tea is another daily source of antioxidants for most Icelanders. They love their English Breakfast or Earl Grey tea and are always boiling a pot. Interestingly enough, there are a number of studies linking high rates of black or green tea drinking to low rates of depression. This same effect was not found with coffee, so it seems that the active ingredient is not the caffeine. While many nutrients in the tea might give a sense of well-being, it is likely that the high concentration of powerful antioxidants (polyphenols) plays a role.

Other unlikely antioxidant sources in the Icelandic diet include the seaweed or dulse that is picked along the beaches at high tide and cooked into bread or soups, the cabbage that is served sweet and sour and hardly looks like a vegetable, the wild sorrel and thyme that grow outside in summer and are added to stews, the dried peas in the very popular split-pea soup, the rye and barley grains that are made into bread or cereals, and on and on. Although they were not eating greens or anything that even vaguely resembled a salad, Icelanders were getting their antioxidants from a variety of sources.

## Good Food Sources of Brain Nutrients

Folate—leafy greens, oranges, peanut butter, beets, and beans

$B_6$—meat/poultry, fish, beans, whole grains, avocados, and bananas

$B_{12}$ (only animal products)—dairy, eggs, and meat/poultry

Thiamin—pork, poultry, red meat, beans, whole grains, seeds, and corn

Chromium—broccoli, mushrooms, beans, seeds, brewer's yeast, oysters, and potatoes

Selenium—poultry, fish, whole grains, seeds, and eggs

Magnesium—whole grains, potatoes, fish, spinach, milk, and beans

Iron—red meat,* poultry,* spinach, cabbage, leafy greens, and beans (*best absorbed)

## Spuds: Friend or Foe?

After fish, potatoes are universally regarded as the most popular Icelandic food. While I generally think of potatoes as a starch rather than a vegetable, many Icelanders I spoke with listed them as their favorite vegetable. Most homes, even in the urban areas, have their own little potato patch, and the spuds seem to grow in even the most inhospitable parts of the island. Traditionally, Icelanders boil up a pot's-worth several times a week to be eaten, hot or cold, alongside their main course. Christine, being a lover of starchy foods, was delighted to hear about the popularity of the potato. She understood why Icelanders considered them to be a tasty accompaniment to every meal and was more than willing to incorporate them into her antidepressant diet. I, however, was not so sure how the potatoes related to the low rates of depression. Were they beneficial? Did they do nothing? Or were Icelanders dodging depression despite the potatoes?

My nutrition education had me believing that potatoes were fast-release, starchy carbohydrates with very little nutritional value. In fact, I thought of them as no better than fluffy white bread in their ability to rapidly raise blood-sugar levels and eventually cause insulin resistance. For Christine, a fast-release food could increase her risk of diabetes and heart disease and, perhaps more importantly, it could aggravate her depression.

## The Vicious Circle

As we know, depression can trigger an unhealthy appetite. Eating foods such as fast-release carbohydrates leads to inflammation and insulin resistence—this, in turn, can exacerbate a depression.

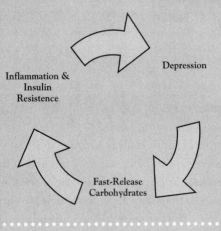

Why would insulin resistance have anything to do with depression? At first glance these two issues might seem unrelated. But let us look closer. As we already know, an abnormal insulin release can cause an increase in inflammation throughout the body. Newer research indicates that this same inflammation may lead to an imbalance in the hormones and neurotransmitters (such as serotonin) that regulate our mood. Given that insulin resistance and inflammation are both linked to depression as well as chronic diseases such as heart disease and diabetes, some scientists hypothesize that depression might be one of the first outward signs of these other health problems.

## Spuds Redeemed

Richard Wurtman, MD, an endocrinologist and depression researcher at the Massachusetts Institute of Technology, gave me a new perspective on potatoes. He believes that these tubers may have some redeeming qualities— especially when it comes to improving our mood. Richard and his wife, Judith Wurtman, PhD, have been studying the effect of carbohydrates on mood since the mid-1960s and are considered to be pioneers in the field. What they have found is that carbohydrates can increase the body's production of natural antidepressants such as serotonin. They believe that this explains why feeling down often results in overwhelming carbohydrate cravings—one is instinctively trying to *self-medicate* by seeking serotonin-producing foods. When I asked Dr. Wurtman to give me some specific examples of foods that could provide a natural antidepressant boost, the first one he mentioned was the potato. He regards it as an excellent example of a filling yet relatively low-calorie food that could help the body make this neurotransmitter.

Richard Wurtman did qualify his findings by saying that this surge in serotonin was only found when the meal had a very high carbohydrate-to-protein ratio (a minimum of 5:1), in other words a pure carbohydrate meal. While potatoes eaten alone could satisfy this criteria, this is not the way they are typically eaten in Iceland. More often than not, they are combined with protein or fat-rich foods that may lessen or even completely block a serotonin release. Therefore, while the occasional potato snack might help boost one's mood, the vast majority of potatoes eaten in Iceland did not seem to be causing a serotonin effect.

David Benton, PhD, a food-mood researcher in Wales, has another explanation for why potatoes might play a role in an antidepressant diet: They taste good. According to his writings, it is the actual sensory experience of the food that gives the antidepressant effect. Eating something that tastes good causes the release of hormones called endorphins that attach to opioid receptors in the brain. This, in turn, gives a general sense of well-being. Most people find potatoes to be exceptionally tasty, and therefore, it is understandable that they create positive feelings and fight depression.

This theory, that carbs have their strongest antidepressant effect by making us happy, would also explain why people who adhere to a high-protein diet tend to become depressed, angry, and tense *and* crave carbohydrates. In general, these diets are less palatable and, therefore, will eventually put one in a foul mood. While this flies in the face of the recently popular high-protein, low-carb diets, in my years of practice, I have noted that people on these high-protein diets have trouble saying "no" to carbohydrates and end up compensating by going on carbohydrate binges. By contrast, people on an intermediate carbohydrate diet (with 60–65 percent of calories from *slow-release carbohydrates*) enjoy their food and tend to feel more positive and energetic.

How about my concern that potatoes are sugar-packed, fast-release foods that can potentially increase insulin resistance? Recently, two separate research groups from Sweden and Canada decided that potatoes warrant a second look. What they found was that declaring potatoes fast release and unfit for human consumption was a bit of a sweeping generalization. In fact, according to their studies, all potatoes are not created equal. The type of potato and the way that it is prepared will make a huge difference in its glycemic index as well as its overall nutritional content. In general, smaller potatoes with a waxy center such as red potatoes or new potatoes have a lower glycemic index than big floury ones. In addition, baking or boiling potatoes in their skins and then cooling them (even if they are later reheated) can lower the glycemic index as much as 40 percent, and eating them whole (rather than mashed) and adding vinegar can also have a beneficial effect on glycemic index. So the Icelandic penchant for using smaller waxy potatoes to make potato salad and their tradition of preboiling their spuds might mean that they are actually eating a slower-release starch. At this point, I was willing to rethink my position on potatoes and even ready to embrace some varieties as part of a carb-lover's antidepressant diet.

## Glycemic Indices of Select Tubers

| Potato Type | Preparation Technique | Glycemic Index |
|---|---|---|
| Waxy | Boiled, cooled, and reheated | 23 |
| Yam | Boiled, unpeeled | 37 |
| Sweet potatoes | Boiled, unpeeled | 44 |
| Waxy | Boiled, peeled or unpeeled | 47–57 |
| Starchy or waxy | Mashed | 74–84 |
| Starchy | Microwaved peeled or unpeeled | 82 |
| Starchy | Boiled, peeled or unpeeled, and served hot | 85–89 |
| Instant (from a box) | Mashed | 85–97 |

## Breakfast: Off to a Happy Start

Christine routinely skipped breakfast because she did not feel hungry. After learning more about the traditional Icelandic diet, she began to question whether this was the best way to start her day. In general, Icelanders have a substantial morning meal. First there is always a cup of tea. Then, on a typical morning they might have whole-grain muesli (my personal favorite was Morgun Gull or "golden morning," which is made from barley and oats and sesame seeds), eggs, or a rye flour pancake. The cereal is usually topped with *skyr* and billberries. On alternate days, breakfast could include dark rye bread, *skyr*, smoked salmon or pickled herring, sliced cucumber, and a hard-boiled egg.

## "Wetting" the Appetite

If you wake up and don't feel hungry, try drinking a glass of water. The mild dehydration that occurs overnight can cause some morning queasiness. I have found that rehydrating with a glass of water can help whet the appetite.

These hearty breakfasts offer another explanation for Icelanders' low rates of depressive illness. In fact, it would be safe to say that breakfast is the most important meal of the day, at least when it comes to mood, focus, and appetite control. But what foods seem to have the most beneficial effect? While fast-release foods such as bagels or white breads quickly raise blood sugar and do offer a short-term improvement in mood and concentration, this effect wears off quickly, and irritability and depression set in. By contrast, the slow-release breakfasts that I was served each morning as I toured Iceland produce a gradual release of glucose throughout the morning. As a result, they ward off fatigue and depression and help maintain a sharper focus. In addition to the mood benefits, eating a slow-release breakfast seems to program us to make better food choices for the rest of the day. For example, in one study, researchers randomly gave school-age children either a slow- or fast-release breakfast. What they found was that those who had been served a slow-release breakfast selected a nutritious, low-calorie lunch from the cafeteria, whereas kids who were given sweetened cereals preferred a sugary, high-fat diet for the rest of the day. This same effect has also been noted in adults. When it comes to treating or preventing depression, it is important not to skip breakfast.

## Slow-Release Icelandic Breakfasts

- Whole-grain cereals with milk or plain yogurt, topped with fruit. Label tip: Choose cereals such as muesli with no more than 5 grams of sugar and at least 5 grams of fiber per 1-cup serving
- Bowl of berries and plain yogurt
- Scrambled eggs with fresh herbs
- Steel-cut oatmeal (not instant) cooked with cinnamon and apples, drizzled with butter or cream
- Whole-grain pumpernickel or rye bread topped with tomatoes, cucumbers, cream cheese, and smoked salmon
- Whole-Grain Rye Pancake (see recipe, page 277)

## Chocolate Bliss

With all the chocolatiers that line the streets of Reykjavik, I could not help but wonder what role this sweet played in Icelandic happiness. After all, I am very aware of the intense feeling of well-being that I can get from a couple of squares of the stuff. Looking at the chemical composition of pure cocoa powder, one finds a number of ingredients that should have a drug-like effect on the brain. There is theobromine—similar to caffeine—that is known to cause a sense of happiness and arousal, biogenic amines (PEA phenylethylamine) that have an effect similar to amphetamine, and finally, anadamide (literally meaning "internal bliss"), that acts similarly to marijuana.

But if these substances were the source of the "feel-good" that we get from eating chocolate, shouldn't we then be as happy after a tablespoon of unsweetened cocoa powder as we are after eating a couple squares of a semisweet, creamy chocolate bar? Clearly this is not the case as most of us are far more satisfied by the bar than the powder. It would appear that David Benton's theory of palatability also explains the wonderful mood-enhancing effect of chocolate.

## Alcohol: A Depressant and Appetite Stimulator

Icelanders do like their Viking beer and local Brennivin schnapps, and I was impressed by the amount of weekend drinking that goes on during the bright summer nights. Nonetheless, I do not recommend this as part of an antidepressant diet since alcohol has been shown to have a profound depressant effect. Furthermore, drinking alcohol causes weight gain by directly stimulating appetite and by providing empty calories. In general our brain and stomach don't recognize the calories from alcohol as filling food, and therefore, we tend to take in the same amount of food calories irrespective of how much alcohol we have consumed.

## Food for the Summer Fog Blues

Returning from Iceland, my plane circled down through a predictably foggy San Francisco sky. It struck me that with such bad weather, it was a wonder that everyone in my hometown did not need to go on antidepressants for the summer months. But later that evening, as I was finishing off a "welcome home" meal of fresh wild-caught Pacific salmon followed by locally grown ollalaberries, I was suddenly reminded that San Franciscans were also on an antidepressant diet. Here on my plate was a medley of omega-3s and antioxidants similar to the Icelandic foods. Thank goodness that salmon and berry season corresponded with the gloomiest part of the year. Nature truly works in wondrous ways, giving us our medicine just when we need it most.

## Christine Does the Hard Work

When Christine reviewed the Icelandic recipes, she immediately voiced her approval. She liked the fact that lettuce and all other leafy greens were absent from the dishes, and she was also drawn to the cuisine's Scandinavian influence since she always enjoyed the meals prepared by her own Danish American grandmother. For the first time in her life, she began to cook for herself and her family on a consistent basis.

The next time that I saw Christine it was late November. She told me that usually by Thanksgiving time she would feel herself sinking into a deep depression. However, this year her mood was holding steady. For months now, she had been eating slow-release breakfasts, and her lunches and dinners included foods like dark rye bread, barley, waxy potatoes, sweet and sour cabbage, beets, plenty of fish and free-range poultry, berries with organic yogurt, and of course, the occasional piece of dark chocolate. She was enjoying her newfound diet and found that being a "dough dog" and a vegetable hater was not such an insurmountable obstacle after all. Although she still had the occasional carb craving, they were much less frequent. She had also started exercising most days of the week and had lost over ten pounds. Unlike other times when she had lost weight, this time she was keeping it off. She was still on an antidepressant, but she felt it was working better and longer because of the other changes that she had made.

At this point, it was very clear to her how closely her mood and her diet are connected, and she lamented the years when she resisted making changes. Even so, she understood why these changes felt so daunting. "I hate to say it," she said with a sigh. "But sometimes you really have to get desperate or scared before you are willing to do the hard work."

# Three Steps to Controlling Depression the Icelandic Way

### Step 1 (Basic):

- Be an omega-3 sponge. Aim to eat *two* servings of low-mercury fish per week or 1 teaspoon of fish oil per day. If fish is not an option, get your omega-3s from walnuts, flaxseed meal, wild greens like purslane (a green often available at farmers' markets), and flaxseed oil.
- Get your brain foods. Even if you dislike most vegetables, find creative ways to get antioxidants, folate, and other B vitamins into your diet. The Icelanders get them from wild berries, a handful of antioxidant-rich vegetables, whole grains such as barley and rye, and black tea.
- Start the day right. Eat a slow-release carbohydrate breakfast.
- Do some form of aerobic exercise for at least thirty minutes, four days a week. Icelanders don't necessarily have to go to the gym to burn calories. They move in the summer while fishing and picking wild bilberries. Despite the inhospitable growing conditions, many Icelanders also keep small gardens. Not only is gardening a wonderful exercise, working in the earth is also an excellent remedy for a bad mood.
- Cook at least one recipe per week from the Iceland recipe section.

### Add Step 2 (Intermediate):

- Create meals that include slow-release carbohydrates (such as whole grains, beans, and waxy potatoes) for an even effect on blood sugars.
- Be even more creative with your antioxidant food sources: think cabbage and more blueberries.

- Cut down on alcohol as this can worsen your depression.
- Tea time. Try to drink 2–3 cups of black or green tea per day. This has been shown to have an antidepressant effect.
- Live a little. Make sure that you choose foods that are pleasurable and give you positive associations—just make sure not to overindulge.
- Do some form of aerobic exercise for at least thirty minutes, five days a week
- Cook at least two recipes per week from the Iceland recipe section.

### Add Step 3 (Advanced):
- Get more omega-3s. Look for a variety of sources, including organic dairy products, Swiss Alpine cheese, omega-3 enriched eggs, wild game, and organ meat from free-range animals.
- For a midafternoon pick-me-up snack, choose a nutritious, pure carbohydrate like a boiled potato salad with salt and vinegar or a baked potato rather than candy, chips, or baked goods.
- Connect to your community. Being a part of a community will do wonders for your sense of well-being.
- Exercise for thirty minutes at least six days a week. Exercise is an excellent antidepressant!
- Cook at least three recipes per week from the Iceland recipe section.

# 8. Cameroon, West Africa: A Cold Spot for Bowel Trouble

Unlike the other destinations in this book, I found myself in Cameroon long before I had a formal interest in cold spots. It was not until years later, when my patient Lynelle asked me to find her a diet to help prevent colon cancer, that I finally made the connection: the area of Cameroon where I had worked as a community health trainer also happened to be a perfect cold spot for colon cancer.

Colon cancer is yet another disease of modernity and industrialization and is on the rise in the United States and around the world. After lung cancer, it is the second most common cause of cancer deaths in the United States, affecting people of all ethnicities. Interestingly, in parts of the world with high rates of colon cancer, we see similarly high rates of a host of other intestinal problems, including hemorrhoids, diverticulosis (outpouchings and infections in the colon), appendicitis, ulcerative colitis, Crohn's disease, peptic ulcers, hiatal hernias, and esophageal reflux. This has led many gastroenterologists to conclude that the same dietary factors that are linked to an increase in colon cancer are also causing these other problems.

In this chapter, Lynelle learns that much of western and southern Africa is a cold spot for colon cancer. She then poses a very important question: why is it that so many of her African American relatives have suffered from colon cancer while the same disease still remains a rarity in her ancestral homeland of western Africa? Inspired by this question, I reexplore the traditional Cameroonian diet and discover the "Five Fs," five main factors that can explain this region's cold spot status. As a part of this search, I revisit the work of an Irish surgeon (known in some circles as the "Bran Man") who did his groundbreaking colon cancer research in Africa almost forty years ago. I also interview one of his modern counterparts—a gastroenterologist who has come up with some interesting new explanations for why many parts of Africa still have low rates of colon disease. Finally, a retired school teacher, a culinary historian, and a chef help me bring the discoveries home by showing me how closely *real* African American soul cooking is connected to its West African roots.

## Meet Lynelle

*"It's really not too bad," I assured Lynelle. "The worst part is the day before when you have to stay near the toilet: the dreaded colon cleanout. The procedure itself is pretty quick and painless." Easy for me to say. Being in my early forties, I have not yet had my own screening colonoscopy. However, most of my patients seem pleasantly surprised by the ease with which they undergo this universally dreaded procedure, and I find myself sharing their reassurances with others.*

*This was the first time I had seen Lynelle in over a year. Being a mother of two school-age boys, a full-time guidance counselor, and a board member of her church left her little time to go to the doctor. Invariably, when she did come in for a visit, she seemed less interested in her own health issues than in asking after my well-being or getting medical advice for her loved ones or fellow church members. Whenever I would question her about her health, she would reassure me that she had an extra dose of her father's "good" genes and would be just fine. A retired school teacher, he was still going strong in his late eighties.*

*Lynelle's most recent visit started no differently than the others. She whirled into my exam room and began by asking about all my family members, remembering each by name. After the usual small talk, however, her tone became uncharacteristically sober, and she told me that she had some serious family worries. These concerned her father and a cousin on her mother's side. Within one month,* they had both been diagnosed with colon cancer. Her father's was early and might be able to be removed, but the cousin's cancer had spread well beyond the colon and the prognosis was very grim. She also reminded me that this cousin's mother (her aunt) had died in her fifties from metastatic cancer of the colon.

Given her family history, I recommended that Lynelle get a screening colonoscopy to look for early signs of colon cancer. I explained that while colonoscopies are the best screening tool we have for colon cancer, they are what we call secondary prevention. That means that they can usually catch the disease in the form of a precancerous polyp, before it grows into an invasive cancer—but they do not prevent the cancer or polyp from occurring in the first place.

## Colon Cancer: How and When to Screen?

While colonoscopy is one of the most effective ways to screen for colon cancer, there are a number of screening options. Beginning at age fifty, you should consult with your physician to find out which screening test is most appropriate for you.

Please talk to your doctor about starting colorectal cancer screening earlier and/or undergoing screening more often if you have any of the following colon cancer risk factors:

- A personal history of colon or rectal cancer or adenomatous polyps
- A strong family history of colorectal cancer or polyps (cancer or polyps in a first-degree relative—parent, sibling, or child—diagnosed at any age or in a second-degree relative diagnosed before age sixty)
- A personal history of chronic inflammatory bowel disease
- A family history of hereditary colorectal cancer syndrome (familial adenomatous polyposis or hereditary nonpolyposis colon cancer)

After some consideration, Lynelle agreed to get the colonoscopy but wondered what she could do for primary prevention. In other words, what would keep polyps from growing in her intestine? She began to tell me about a magazine article that she had read recently on this very same subject. The article discussed how colon cancer is on the rise in the African American community. It went on to point the finger at cigarette smoking, junk foods, and greasy Southern cooking (which the article described as "killer soul") as being the major reasons for this alarming statistic. Lynelle felt good about the fact that she did not smoke and rarely ate soul food, but she was not completely buying this explanation. After all, no one in her family smoked, and her aunt and cousins, who lived on Long Island, never ate chitlins, ham hocks, or other foods that she felt would qualify as "killer soul." Instead they ate a standard American diet of mashed potatoes, rice, baked chicken, spaghetti, pizza, steak, and sides of frozen vegetables. Obviously none of them were the least bit protected from colon cancer.

Lynelle's father, on the other hand, was raised in rural South Carolina. He had spent much of his early life eating something that Lynelle thought was much more akin to soul food, and yet he had stayed healthy and vigorous until his mideighties. And even now that he had been diagnosed with colon cancer, it was a relatively small tumor and had come on late in life. Lynelle wanted to know what she could do to ward off her own colon cancer. In our five years of working together, this was the first time that Lynelle had asked me for advice about herself. I was determined to give her a good answer.

Telling Lynelle about my interest in modern disease cold spots, I mentioned that western and southern Africa have some of the lowest rates worldwide for colon cancer and for other colon problems. "How strange," she commented, "that African Americans have so much colon cancer and Africans have so little. After all, many of my ancestors came from Africa, somewhere. These big differences in colon cancer in just a few generations must mean that food is really important. Probably much more important than genes."

## Dr. Burkitt's Potty Talk

From what I can tell, Denis Burkitt, MD, would have heartily agreed with Lynelle's observations about the powerful connection between diet and colon cancer. In fact, he is one of the first researchers to have ever discussed the link between food and colon disease in the medical literature. I was first introduced to Denis Burkitt's work in a dusty library located in the main medical school in Yaounde, Cameroon. I had been in the country for several months when a professor at the university was kind enough to let me use the library to get information for a health-worker training manual. Among the hundreds of books in French about infectious disease, a book written in English caught my eye. Entitled *Western Diseases: Their Emergence and Prevention*, the book, which was published in 1981, was edited by two physicians, Hugh Trowell and Denis Burkitt. The name Burkitt immediately rang a bell because of Burkitt's lymphoma, a disease that this very same doctor had first described in an African boy. However, in this book, he was talking about something quite different. Each chapter focused on a specific modern chronic disease and discussed what was, in effect, a cold spot for that disease: a place somewhere in the world where people still had very low rates of that particular health problem. One of the first chapters, written by Burkitt himself, was about diseases of the colon.

A native Irishman, Burkitt attended medical school in Dublin, and then in the late 1940s enlisted in Colonial Medical Service as a general surgeon. His first assignment, which lasted for well over a decade, was a hundred-bed hospital in rural Uganda. While working there, he was struck by the fact that he hardly ever operated on what he called "noninfective" diseases of the bowel such as colon cancer, hernias, hemorrhoids, and diverticulosis. (These diseases are so common in the West that general surgeons often refer to them as their "bread and butter.") Burkitt wondered why these bowel problems were practically nonexistant in Uganda and decided to make it the focus of his research. Throughout the 1960s, he traveled around western, central, and southern Africa, collecting data and interviewing other physicians. In the end, he made a number of important observations that he would have been hard-pressed to discuss in polite company.

First of all, he noticed that white Europeans seemed to take twice as long to digest and pass their food as rural black Africans. (This was based on how long it took for a swallowed pellet to show up at the other end.) Second, he described European stools as being "hard and faceted and malodorous" while black Africans had softer and less foul-smelling bowel movements. Third, he discovered that the actual bacteria growing in the intestines of most African blacks was a much healthier form of bacteria than the bacterial colonies that he found in urban whites. Finally, Burkitt observed that when black Africans moved from their rural villages to more urban areas, they seemed to lose their protection from colon cancer and other colon problems. As a matter of fact, their rates of colon disease slowly began to match those seen in Europeans and North Americans. This sudden shift made him conclude that traditional African foods could help prevent colon problems, whereas a modern diet seemed to trigger them. Speaking about these issues to medical colleagues, he would often say, "The only way we are going to reduce [this] disease is to go backward to the diets and lifestyles of our ancestors."

## African Foods: Beyond the Stereotypes

Lynelle confessed that although she loved African cloth and African music, she really did not know what to think about traditional African food. She worked long hours, and it felt like a scramble to get a hot dinner on the table before her kids filled up on snacks from the cupboard. She needed to cook things that were simple yet would appeal to her sons. Her only experience tasting African cuisine was at a local international food fair, and she remembered it as being too spicy and greasy for their taste buds. She was also certain that the ingredients would be hard to find in the local market.

Lynelle's unfamiliarity with foods from Africa did not surprise me. Even my foodie and chef friends who are well versed in the regional variations in European, Asian, or Latin American cuisine are often clueless about African foods—especially from areas south of the Sahara. The continent remains lumped together, in many of our minds, as one amorphous place, almost as if it were one big country rather than a collection of nations and distinct geographic zones. So few of us know how the taste of East African meals differs from those in the west, the north, the center, or the south, much less what recipes are particular to one area of any given African country. Even a city as cosmopolitan as San Francisco has only five or six North African restaurants and even fewer specializing in regional sub-Saharan foods.

Unfortunately, African cuisine is also subject to some erroneous notions. In one instance I tried to interest a patient in a couple of Cameroonian recipes only to be told that it was "hard to get excited about famine food." Although misguided, this response is understandable given that the most common images that we see of Africa in the media are desert wastelands and clutches of children with spindly limbs and bellies swollen by malnutrition. What this patient did not understand, however, was that indigenous African cuisine did not *cause* this famine and suffering. In fact, it is the other way around. Much of the chronic and pervasive famine that we believe to be a fixture in countries such as Mali, Kenya, Ethiopia, and Sudan is caused by a *loss* of indigenous diets.

## The Seeds of Starvation: A Brief History

Until the beginning of the twentieth century, much of sub-Saharan Africa was inhabited by nomadic farmers and hunter-gatherers who survived on wild foods or drought-tolerant crops of indigenous plants. Their herds of livestock were just big enough to help work the fields and put meat on the table without being so plentiful that they would overgraze. Even people living in the more barren areas could usually gather or grow an adequate supply of food year round. Then, as parts of the continent were colonized by Europeans and other African tribes, most nomadic or hunter-gatherer tribes became confined to small plots of nutrient-depleted, overused land. Many of their farming and gathering areas were taken over by large-scale agriculture, and whole countries began to produce one or two high-yield crops rather than the dozens of indigenous varieties that were previously cultivated. As a consequence, many of the native seeds and nutritious indigenous foods were lost, leaving millions of Africans to suffer from malnutrition.

These vast stretches of overfarmed and deforested land have also caused climatic changes, leading to even more devastation. Huge shifts in rainfall have turned natural deserts into flood plains and lakes into deserts. With these changes, famine, which used to be seasonal and limited to small areas, now affects millions of people throughout the continent. Rather than helping, food corporations such as Coca-Cola and Nestle have made the situation even worse. Recognizing Africa as a wide-open market, they have flooded the rural areas with low-nutrient, sweet, quick-calorie products that rapidly replaced any surviving indigenous foods.

## Ntui

What the media rarely tell us is that there are some parts of sub-Saharan Africa that still have diverse ecosystems and that are home to some of the most delicious and complex cuisines on this planet. One such area that I am personally familiar with is Ntui, a small village in the rainforest of central Cameroon. This was my home for the year I spent working for Save the Children Federation as a rural health trainer. At that time I did not have a formal interest in cold spots; in retrospect, however, I realize that I was living solidly in a colon disease cold spot zone.

Fortunately Ntui has not changed very much since I was first there in the mid-1980s. According to medical colleagues who have been there recently, the mayor's house and the bar now have electricity, a couple of the streets are paved, and the hospital has a few more walls. However, the most vital parts of village life remain the same. On any given day, the smell of simmering peanut stews, fresh grilled chicken, and plantains still wafts out from the courtyard of every mud compound. There is still a bustling Tuesday and Saturday market overflowing with locally grown fruits and vegetables of all varieties. The *porc-épic* (porcupine) remains a local delicacy, and no one can host a funeral or a wedding without spending days preparing a multicourse feast. But what surprised me the most was to hear that Madame, one of the most memorable women in the village, was still serving up the same wonderful foods that she has for decades.

## Madame's *Chantier*

Madame most certainly has a first name and a last name, but no one ever calls her by them. For anyone who eats at her establishment, she is simply referred to as "Madame" (French for "Mrs."). She runs what Cameroonians call a *chantier*, a small, informal restaurant inside her home. During the time that I knew Madame, her kitchen consisted of a gas camping stove, a fire pit lined by a circle of rocks, and a rickety wooden table where she stored her battered set of tin pots along with piles of fresh greens, onions, root vegetables, and spices. The dining area was a kerosene-lit mud room with a couple of low benches and tables that matched the one in the kitchen. On most evenings my dining companions were a handful of other villagers, a lone lorrie driver passing through town with cargo from the north of the country, and, of course, a gaggle of Madame's children, nieces, and nephews.

While I spent my days working on local community health projects, at quitting time I would shower off the dust of the day and run over to Madame's *chantier* to watch her cook. It was in her rudimentary kitchen that I developed what has turned out to be a life-long love for cooking and eating indigenous foods. My journal, which I kept religiously during my stay in Cameroon, reads more like a cookbook than a travel diary, with each new entry chronicling a different meal: grilled fish or free-range chicken seasoned with exotic spices, peppery goat stews with a rich peanut sauce served with manioc root, dozens of types of wild greens often prepared with a nutty tasting brown rice, and plantains boiled or grilled. Then there was the wild game. And when I say wild, I mean *wild*. Madame would cook up whatever was caught in the bush that morning: porc-épic, boar, antelope, boa constrictor, and bush monkey were regular fare at her chantier. And finally, there was the *ndole* (*N-do-lay*). To this day, ndole remains one of my favorite meals, and I still remember every detail associated with my first sampling of this most unique dish.

# Discovering *Ndole*

I had barely been in Yaounde (Cameroon's capital city) for twenty-four hours when my coworkers loaded me into the back seat of a jeep and headed out to Ntui. As we entered farther and farther into the dense jungle, bumping our way along endless miles of rutted dirt roads, I found myself wondering: was it really such a great idea to take this job thousands of miles away from home, deep in the middle of a rainforest? This was a place where very few urban Cameroonians had ever ventured, much less foreigners.

Finally, after what seemed like an eternity, we parked on a muddy embankment and piled out. It was a moonless night, and all that I could see was a warm light coming through the cracks of a house nearby. I remember following my hosts into what turned out to be a small mud hut with a corrugated roof. Inside were half a dozen locals laughing, drinking beer, and scooping a thick green stew out of tin bowls. Cameroonian Macosa music pumped out from a huge boombox propped in the corner. The room smelled intoxicatingly good—but foreign. One of my new colleagues named Pierre said to me, "You have not arrived in Cameroon until you taste your first ndole."

Jet-lagged, my gut throbbing with anxiety, I sat down at the wobbly wooden table to eat. Suddenly Madame appeared from the back room, a tall, elegant woman wearing a beautiful purple boo-boo (traditional dress), her hair wrapped up in an equally ornate cloth. In a mixture of French and pidgin English, she greeted me as if it were an everyday event that a dusty, travel-weary, white woman appeared in her chantier: "*A bonsoir, you dun come. S'il vous plait shiddown fa chop.*" (Good evening, you have arrived; please sit down to eat.)

Then she brought out the ndole. At first I hesitated, wondering if I would be making a serious mistake by sampling this black-green stew served with a sprinkling of peanuts and dried shrimp. I watched my colleagues take pieces of what looked like a massive chunk of string cheese (in actuality it was boiled and braided manioc root) and dip it in the ndole. My hunger got the best of me, and I bravely followed suit.

When the ndole first hit my tongue, I was shocked that such a complex and delicious flavor could be found in a pile of mush. It turns out that I was eating a perfectly slow-cooked combination of dried bitterleaf, shrimp, beef, roasted peanuts, onions, tomatoes, and a litany of spices, including ginger and garlic. After several bites, any anxiety I had about my new surroundings seemed to vanish. How could I not be in a wonderful place when the food tasted so good? Indeed, in the months that followed, whenever I needed a lift, I would go to Madame's and put in a special request for her ndole.

## Gathering the Clues: The Five Fs of Colon Cancer Prevention

So what was it about Madame's dishes, including the ndole, that was especially protective against colon cancer and other colon diseases? Denis Burkitt and his research successors posed a similar question. In their search for an answer, they compared foods that were eaten in traditional South and West African communities with those eaten in urban areas. Not surprisingly, they could not isolate one specific dietary factor. However, as I reviewed their writings, I realized that there were five main themes that emerged. These I have called the five Fs of colon cancer prevention: fiber, less flesh, fermented foods, foraged foods, and select fats.

## F #1: Fiber

The surgeon Denis Burkitt noted early on during his time working in Africa that people living in the urban areas ate a lot of foods made with processed, low-fiber wheat flour. By contrast, traditional communities relied more on high-fiber, traditional starches such as millet, sorghum, teff, local hybrids of maize, plantain, and brown rice (which, interestingly enough, is actually indigenous to West Africa). Comparing the urban and rural diets, Burkitt concluded that fiber, a nonabsorbable carbohydrate, was playing a key role in protecting the lining of the colon from injury and disease.

While this conclusion may have been revolutionary in the late 1960s and early 1970s, at this point most experts would agree that a high-fiber diet helps to prevent or treat colon problems such as constipation, hemorrhoids, irritable bowel disease, or diverticulitis (infections that occur in abnormal outpouchings in the intestinal wall). Certainly most of the research shows that there is a clear link between more fiber and lower rates of these health problems. But what about colon cancer?

Everything I read about fiber and colon cancer seems to offer different (and often contradictory) information: "It helps. It does nothing. Eat flax. Eat fruits. Eat oat bran." I have sorted all the fiber and colon cancer studies into three impressive stacks on my desk. In one are all the studies that show a protective effect of fiber; in another equally large pile are the studies that refute this connection; and the third holds all the "inconclusives." Like so many aspects of nutrition research, these studies are fraught with problems. First of all, most studies are too short term to really see an effect. It can take a colon polyp years to grow into a cancer; therefore, following patients on a high-fiber diet for a mere ten years (or less) is simply not adequate. Second, most studies lump soluble and insoluble fiber into one big category. It turns out that soluble fiber (the kind you get from fruits and beans) is very good for you but probably does not do a huge amount to protect you from colon cancer. Insoluble fiber from grains, which can act like a scouring pad in our intestines, seems to do a lot.

I am especially impressed by the biggest fiber colon cancer trial to date, which studied over 500,000 people in ten European countries. It showed that people who ate more than 30 grams of fiber per day (soluble plus insoluble) had approximately half the risk of colon cancer as compared with those who ate 12–15 grams of fiber daily. Furthermore, those who had followed this regimen the longest had the most protection. Based on this information, I told Lynelle to try and get at least 25 grams per day of fiber. I also encouraged her to start giving her kids a high-fiber diet now to prevent health problems in the future.

# A Day's Worth of Fiber

|  | Approximate grams of fiber |
| --- | --- |
| Beans of any variety (¾ cup cooked) | 12 |
| Avocado (¼ cup mashed) | 4 |
| Greens (collard, mustard, etc) (1 cup cooked) | 4 |
| Okra (½ cup cooked) | 2 |
| Brown rice, Wild rice (½ cup cooked) or Millet (1 cup cooked) | 2 |
| Sweet potato (1 medium) or Plantain (1 medium) cooked | 4 |
| Peanuts or Sunflower seeds (roasted, 1 ounce) | 2 |
| Total | 30 |

Source: USDA Nutrient Data Laboratory (www.ars.usda.gov)

Stephen O'Keefe, MD, a gastroenterologist who has spent part of his career working in South Africa, has yet another idea about why colon problems are such a rarity in many parts of rural western and southern Africa. Talking to me from his office at the University of Pittsburgh Medical School, he told me about his time spent working in a public hospital in Cape Town in the 1980s. "Despite doing a lot of colonoscopies, I rarely found polyps, colon cancer, or other colon health problems. It was clear to me that there was something in the diet." However, he thinks that "something" is more likely to be a low-meat diet rather than a high-fiber one. He explained to me that even when rural blacks move to urban areas and start eating more refined carbohydrates, their colon cancer rates do not match those found in the white populations. "It is not until black South Africans become solidly middle class and start eating a lot of pork and red meat that they really start getting colon cancer."

## Meat and Colon Cancer

The following are linked to increased risk of colon cancer:

- Eating red meat more than once a week
- Eating grilled meat more than twice a week
- Eating processed meats (nitrate-treated, smoked, or cured)

Note: There is evidence to suggest that regular poultry and fish intake is actually *protective* against colorectal cancer.

In Madame's chantier, meat was rarely used as a main feature in a meal. Similar to what I discovered in the other cold spots, animal products in Ntui were expensive or hard to come by and therefore treated as special commodities. One of the only times that I was served a true slab of meat was during a holiday or other special celebration. Indeed, the most delicious dishes that I sampled were completely vegetarian, using grains and legumes such as peanuts and black-eyed peas as the main protein, or else they were stews containing some meat plus a large proportion of unrefined grains and vegetables.

Why are these big servings of meat associated with colon cancer? There probably is not one single explanation but rather a number of contributing factors. First of all, the way meat is *processed* may play a major role in stimulating the growth of cancerous polyps. There is convincing evidence that the chemicals used to smoke or preserve meat may in fact be among the culprits. Therefore, when shopping for meat, try, whenever possible, to buy fresh cuts and pass over the prepackaged or deli case items that contain any additives; these are mostly found in hot dogs, cold cuts, bacon, jerkies, and pepperoni. Specific ingredients to watch out for include sodium nitrate (saltpeter), sodium nitrite, N-nitrosodimethylamine, sodium phosphate, and smoke flavor.

Secondly, the way meat is *cooked* may also be significant. We know that cooking meats at very high temperatures—like grilling or barbecuing—produces heterocyclic amines (HCAs), which are cancer-causing compounds. HCAs form when high heat reacts with proteins in animal flesh. To reduce your exposure to toxins in grilled meats, poultry, and fish, I recommend that you limit your grilled meat to one to two servings per week and follow the steps listed in the sidebar "Safe BBQ 101."

### Safe BBQ 101: Steps to Cut Down on Cancer-Causing HCAs

- Precook meats and poultry in the oven to reduce time on the grill.
- Flip meat frequently while grilling to keep it from getting charred.

- Place meats and poultry far enough from the flames to prevent fire flare-ups.
- Marinate meat, poultry, and fish first. This decreases the amount of HCAs formed. It's important to use thinner (more watery) marinades rather than thick sauces for this trick to work.
- Grill lots of fruits and vegetables. These don't form HCAs when heated.
- Add antioxidant-rich fruits and vegetables like garlic, onion, and even berries into meats and marinades—research shows that these foods reduce the amount of HCAs.

## F #3: Fermented Foods

While a low-meat diet may expose the gut to fewer chemicals and toxins, Stephen O'Keefe believes that the real link between less flesh and the lower rates of colon disease can be explained by "bugs." The human intestine is filled with bacteria. But different bacteria can have very different effects. While some help maintain a healthy gut lining, others are destructive to the intestinal wall. Study after study shows that heavy meat eaters, as compared to vegetarians and occasional meat eaters, have less healthy bacterial populations in their intestines. Dr. O' Keefe explained that traditional African foods contain high levels of probiotics, nutrients, and bacteria that are beneficial for the gut lining. By contrast, our refined Western diet offers very little of these protective bacteria. To make matters worse, a diet high in refined carbohydrates and meat promotes the growth of unhealthy (or even destructive) bacteria in the intestine.

This connection between bacteria and gut health was certainly not news to me. There are many studies that show that excessive antibiotic use can kill the healthy bacterial populations in one's intestinal lining, leading to diarrhea, indigestion, bloating, and problems with absorbing nutrients. It makes sense that eating the wrong foods can have a similar detrimental effect. What was news to me, however, was the link between gut bacteria and colon cancer. Granted this area of research is still in its infancy. Nonetheless, I found a number of studies published in reputable journals that support the idea that there is connection. O'Keefe is not sure whether the bacteria themselves affect the colon or whether they produce secondary substances that can turn cancer cells on or off. What seems clear is that lactic acid–producing bacteria such as *Lactobacillus* that are found in fermented dairy, grains, and vegetables seem to be the most protective, while the *Bacteroides* species, found in higher concentration in people who eat a high-meat, high–refined carbohydrate diet, seem to be more cancer causing.

## Good Sources of Probiotics (Healthy Gut Bacteria)

In general, my recommendation is have *at least one serving per day* of a whole fermented food rather than choosing functional products or supplements.

**BEST CHOICES:**
- Dairy products (preferably sugar-free, preservative-free) made with *L. acidophilus* and *B. bifidum*. These include yogurt, kefir, cottage cheese, and buttermilk. To ensure maximum survival of the probiotics, use before the expiration date. Look for the "Live and Active Cultures" seal.
- Nondairy sources, including traditionally fermented fruits, vegetables, and grains (kim-chee, sauerkraut, pickled relish, miso, tempeh, *natto*, soy sauce, *ogi*, *mahewu*, etc.). Make sure that your pickled foods are prepared with salt brine, not vinegar, to ensure they are a source of good bacteria.

**OTHER OPTIONS:**
- Probiotic supplements. According to *Consumer Reports*, most plain yogurt has much more beneficial bacteria per serving than the probiotic pills they tested.

In the region of Mpumulanga in South Africa, where Dr. O'Keefe and his research team continue to work, sour milk is one of the standard drinks. Sour milk is made, as the name implies, by letting fresh whole milk stand out on the front stoop for several days until it becomes a slightly tangy liquid yogurt. While this might sound distasteful to those of us who would never dream of leaving a milk product unrefrigerated, remember that this is pretty much the same process that is used to make yogurt. Cultured sour milk and yogurt are both good sources of those healthful colon-protecting bacteria (or probiotics). Other fermented foods found throughout Africa, including relishes or pickled foods and fermented corn, manioc, and millet are also excellent sources of beneficial bacteria.

## Good Sources of Prebiotics

Prebiotics (also called fructo-oligosaccharides and inulin) are fibers that feed the good bacteria in the gut. Eating foods containing prebiotics helps raise the total amount of good bacteria in the intestines. Such foods include:

- Jerusalem artichokes
- Onions and leeks
- Bananas
- Wheat
- Chicory root
- Tomatoes

Although many of us get unbearably gassy when we eat milk products, we tend to do better with fermented dairy foods. This is because the gas-causing lactose (sugar) in fermented products is predigested by the bacteria.

Home-brewed beer was another standard fermented food in Ntui. My neighbor, a local bigwig who owned the village bar, was constantly distilling millet or manioc root beer. He would use the smallest excuse to throw a party so that he could serve up these brews. While I took great delight in most of the flavors I sampled that year, I must confess that these requisite sips of warm, semisweet, musty brew from the calabash were not my favorite moments. Recently I had the pleasure of interviewing James McCann, a historian at Boston University and an expert on African cuisine. When the subject turned to African home-brewed beer, his description of *tella* (the Ethiopian fermented drink made from barley or sorghum) rang true for me: "It tastes like flat Coca-Cola in which someone had put out a cigarette." While this particular concoction is not be for everyone, there are many other home brew recipes that might please your taste buds *and* your intestinal lining. For a list of excellent references for home fermentation projects, see Appendix G.

## F #4: Foraged Foods

Wild greens are a central feature in most traditional diets throughout sub-Saharan Africa. In Ntui during the rainy season, everyone became an expert forager. At night I would lie awake and listen to the rounds of torrential rain pounding the tin roof of my hut. Then, in the morning, I would be greeted by glistening sunshine and the sound of grass rustling under my window. Looking out, I could catch glimpses of children as they wandered through the wet thicket around the hut picking clumps of greenery. Occasionally a child would pop a clump of her harvest into her mouth and munch as she picked. The first time that I witnessed this I asked one of my urban Cameroonian colleagues about the foraging. He dismissed the activity as something that "poor villagers do" and assured me that his wife would never be caught picking wild grasses. However, when I asked Madame about the harvesting of the wild greens, she seemed delighted that I had raised the subject. According to her, the jungle in the rainy season produced the tastiest foods for the village. In fact the wild bitterleaf, or *folon*, that was being picked outside my hut was her favorite green for ndole and other stews. She gave two reasons for this preference: it was more flavorful than the farmed varieties one could buy in the market, and it helped maintain a healthy digestion.

It appears that Madame, with no formal education, already knew what researchers have spent years in the lab figuring out: wild greens are rich in many vitamins and antioxidants that protect the colon from injury. Folate, a B vitamin that is concentrated in wild greens, seems to play an especially important role in colon health. The likely way that this vitamin prevents colon cancer is by a process called "DNA methylation." In other words, folate blocks the reproduction of DNA within cancer cells.

One large twenty-year-long study showed that the risk of colon cancer is decreased by 60 percent in people who eat folate-rich foods daily versus those who rarely eat these foods. Interestingly, these same positive results are not seen in people who just take folic acid, which is the supplement form of folate.

## Getting Enough Folate from Your Indigenous Diet

Most people should be able to meet their daily folate needs by eating whole foods. For adults, at least two servings (1 cup total) of cooked greens plus 1 cup of any of the cooked indigenous starches in Appendix B should do the trick. For children, half this amount is enough. Because folate deficiency is a major cause of neural tube defects, I do recommend that all women who are pregnant or are planning a pregnancy should take a daily prenatal vitamin containing at least 400 mcg of folic acid just to ensure that they are getting enough of this nutrient. The RDA (recommended daily allowance) for folate is 200–300 mcg for children and 400 mcg for nonpregnant adults.

Not only does folate have protective powers when it comes to fighting colon cancer, it might also play a role in decreasing or preventing other inflammatory conditions in the bowel such as Crohn's disease or ulcerative colitis. It does this by converting homocysteine (an amino acid that can be toxic to cells) to less toxic products such as methionine and S-adenosylmethionine (SAM). While this is still very new research, the dual benefit of folate in preventing both colon cancer and other chronic colon problems gives us a good clue as to how these different health problems are related.

## The Alcohol–Folate Colon Cancer Link

Drinking more than two alcoholic drinks per day puts you at a higher risk for colon cancer. Why? No one knows for certain, but folate deficiency may be one of the explanations since alcohol blocks the beneficial effects of folate.

## F #5: Select Fats

A final F relating to colon cancer prevention is fat. Eating a diet high in calories is connected to colon cancers. Since fat offers more calories per ounce than proteins or carbohydrates, it is reasonable to assume that a moderate- to low-fat diet might help prevent this disease. However, the issue is not this straightforward. As we have seen before, there are fats to avoid and others to seek out. Diets high in saturated animal fats seem to promote colon cancer. (Of course, it is hard to say whether it is specifically the fat or just red meat in general that is an issue.) By contrast, diets that are high in monounsaturated and omega-3 fats seem to be protective.

## Omega-3s Yet Again

A study done by a team of South African gastroenterologists underscores the important role that omega-3 fats play in colon cancer prevention. The researchers focused their attention on a group of South African fishermen with exceptionally low rates of colon cancer. After spending some time surveying this group, they discovered that many of the fishermen had rather unhealthy habits: they smoked, ate high-salt foods, and had a relatively low-fiber diet. However, there was one healthy behavior that they all shared: they ate a lot of fish. This fact, concluded the researchers, offers a likely explanation for why they were protected from colon cancer. I report this not to encourage anyone to smoke or eat high-salt, low-fiber foods. It is simply an illustration of how omega-3s, yet again, can be especially powerful medicine.

If we look at the dishes cooked up in Madame's kitchen, we can learn a lot about which fats are beneficial and which are harmful when it comes to maintaining a healthy gut. Madame's dishes were rich in omega-3 fats thanks to the fish, wild game, greens, nuts, and seeds. Furthermore, like most indigenous cooks, she used oils sparingly, and most of the fats in her meals came from whole foods. On the occasions when she did cook with oil, her selection included either unrefined peanut oil or palm fruit oil.

Both oils were kept in earthenware jugs and stored in the coolest corner of her kitchen. I recall watching her occasionally pour out a tablespoon or so to sauté onions as a base for a stew; however, her main use for these oils was as a "finishing oil": a small amount of brick-red palm fruit oil elegantly drizzled over a dish before it was served to her guests.

Interestingly enough, both unrefined peanut and palm fruit oils may play a role in colon cancer prevention. Peanut products (including raw peanut butter) have high concentrations of polyunsaturated fatty acids and an antioxidant called beta-sitosterol. At least in a test tube, this antioxidant has been shown to prevent the replication of cancer cells. Similarly, palm fruit oil is high in cancer-preventing antioxidants such as vitamin A and vitamin E. (Please note: unrefined palm fruit oil is very different from palm kernel oil. While the orangy-red palm fruit oil has anticancer and anticlotting properties, this is not the case with the clear yellowish palm kernel oil.)

## Unrefined African Cooking Oils

| 1 tablespoon | Calories | Total fat grams | Sat fat grams | Mono fat grams | Omega-6 grams | Omega-3 grams | Smoke point (°F) |
|---|---|---|---|---|---|---|---|
| Palm fruit oil | 120 | 14 | 7.0 | 6.0 | 1.0 | 0 | 450 |
| Sesame oil | 120 | 13 | 2.0 | 6.0 | 6.0 | 0 | 250–350 |
| Peanut oil | 120 | 13.5 | 2.3 | 6.2 | 4.3 | 0 | 275–320 |

In your cold spot cooking, you might want to choose peanut, palm fruit, or even sesame oil to ensure that your Cameroonian dishes have an authentic flavor. (An ndole cooked with olive oil just does not taste right.) My personal preference is to use unrefined peanut oil unless you are cooking at a very high heat, in which case palm fruit oil or partially refined peanut oil are the best choices. Both peanut and sesame oil have a delicious flavor, and either one has a better environmental rap than palm fruit oil. (Farmers in Indonesia and Malaysia are cutting down rainforests to start lucrative palm oil plantations.) Regardless of which oil you choose, I would recommend using it in small amounts (no more than one to two teaspoons per serving) since any oil is highly caloric and will do a swift job of adding inches to the waistline.

## What About Deep-Fried Food?

Although some people associate deep-frying with West African cuisine, traditionally this is not an everyday cooking method. Because frying required a powerful heat source and large amounts of refined oil, Madame would reserve it for treats and special occasions. These days, however, refined oil has become affordable and much more plentiful. As a consequence, deep-frying has begun to replace grilling and slow-cooking as one of the most popular cooking techniques in many Cameroonian homes. A deep-fry has the triple benefit of shortening cooking time, preserving cooked meats longer before spoilage (a real plus in a tropical climate), and creating a crispy brown product that appeals to our fat-loving taste buds.

As delicious as deep-fries can be, there are a number of serious downsides to this method of cooking. First of all, it is extremely caloric. For example, a three-ounce serving of roasted chicken breast has approximately 140 calories (5 grams of fat), whereas the same meat deep-fried will give you more than 75 percent more calories and fat! Second, unless the oil is changed between fryings and kept below its smoke point, the high heat will quickly convert the polyunsaturated fats in the oil into harmful cancer-causing compounds. Finally, most oils that have a smoke point high enough for a deep-fry are *refined* oils. These oils are refined by exposing them to high heat and a variety of solvents. This results in a nutrient-depleted product that contains chemical residues—a far cry nutritionally from the unrefined oils that we can use in our low-heat, slow-cooked traditional recipes. Despite the health concerns related to deep-frying, all of us crave the occasional fried treat. My feeling is that there is no reason not to indulge so long as it is for a *rare weekend feast* or other special occasion. In the recipe section, I offer you some of my favorite cold spot deep-fry feast recipes with the caveat that, when you choose to deep-fry, you should follow the tips in the sidebar to minimize health risks and maximize taste. (Also see Appendix C for more information about cooking oils.)

## Deep-Frying Responsibly

- Deep-fry foods that have a natural protective skin (potatoes, chicken with skin, etc.) or a batter coating. This allows the outside to become brown and crispy, while the inside remains moist and tender.
- Use the *lowest* heat possible and never heat your oil to the point that it produces smoke. I stay somewhere around 350°F, and I use a deep-fat frying thermometer to keep track of the temperature. (Contrary to popular belief, it is quite possible to fry at this lower temperature as long as your oil is fresh and you do not overload the pan with food.)
- Do not overcook your food. Deep-frying works by heating the water molecules in the food. If you overcook your food, its natural moisture will be lost and replaced by oil. This is a surefire way to get a greasy product!
- Use an oil with a relatively high saturated-to-unsaturated fat ratio to minimize the formation of toxic compounds. Unfortunately most unrefined native oils do not fit this bill. My personal choice is palm fruit oil, partially refined peanut

oil, or even olive oil, although many chefs I know prefer
canola or safflower oil.

- Discard your oil between fried meals. Research shows that
with each successive meal, reused oil develops exponentially
more harmful oxidized compounds. If you are frying a large
amount of food at one time, discard your oil after every
third to fourth batch.
- When eating out, definitely avoid fried foods because they are
usually made with highly refined—and often partially hydrogenated—cooking oil that is stored and reused in a deep-fryer.

## Ndole, Yogurt, and Millet: A Microcosm of the Five Fs

Most of the recipes I had collected in Ntui were a part of my weekly cooking repertoire: I braised or stewed some kind of seasonal green several times
a week, millet was one of my family's favorite grains (we often had it for
dinner and then breakfast), I made batches of *pili pili* sauce on a regular
basis to use on chicken and fish, and, whenever I could get my hands on
tender pods, I made spicy okra. But as I was preparing to give Lynelle my
favorite Cameroonian recipes, I was troubled by the fact that one notable
dish was absent from the list: I had no recipe for ndole.

This dish, as Madame used to prepare it, had twenty-five steps. My
recollection is that she would start her ndole sessions shortly after the
first rooster crowed, and the stew would *finally* be ready to serve as the last
rays of sun were disappearing over the palm trees. It simply is not possible
for me to invest this kind of time to make any meal—no matter how
delicious the result. I was resigned to the fact that Lynelle would never
sample this memorable dish and that the Cameroon chapter in this book
would have to go to print without an ndole recipe. Then, as luck would
have it, I was gifted an "ndole-lite" that is just as nutritious and almost as
scrumptious.

While giving a lecture about indigenous diets at the local medical school, I happened to mention that I loved ndole. After the talk, a nursing student named Genevieve came up and introduced herself. She was from Douala, on the west coast of Cameroon, and she offered me *her* family version of *dolait* (as it is pronounced in Douala), which she was certain would be a more succinct version than Madame's. The next week she appeared in my office with the recipe along with a box of ingredients! I brought the ndole-in-a-box home, followed the easy recipe, and within thirty minutes, I had ndole.

That evening for dinner, I served the dish with a side of plain yogurt and toasted millet. (The yogurt is my replacement for ogi, the traditional West African fermented millet since I am still not brave enough to ferment my own millet.) A true ndole connoisseur might be able to tell the difference between the thirty-minute and twelve-hour version, but for the rest of us, Genevieve's recipe is just perfect. As my family was polishing off the last scoop, it suddenly occurred to me how critical it was to include ndole in this chapter. In addition to being delicious, when served with millet and a fermented food such as yogurt, it is also an all-in-one example of the five Fs. There was plenty of *fiber* from the onions, greens, peanuts, and millet, and a rich supply of folate and other nutrients from the *foraged* greens; the seafood, greens, and pumpkin seeds (if used) offered the healthy omega-3 *fats*; and the meat (*flesh*) is used sparingly. The *fermented* yogurt supplied the good bacteria needed to maintain a healthy intestinal lining and improve the nutrient absorption from the millet. Finally, the onions and garlic were excellent prebiotics, serving to promote the growth of these beneficial bacteria. How could I possibly think that I could not include an ndole recipe? This dish was a perfect microcosm of the West African colon protection diet.

## Closing the Loop

*Lynelle called me to tell me that she had tried the recipes for ndole, okra, pili pili chicken, and collard greens. All four dishes were easy to make, and to her great surprise, had received unanimous approval from her family. She also told me that she had recently had an interesting conversation with her father, and* apparently she had the wrong idea about the foods that he was eating in his younger years.

He grew up on a farm in South Carolina, and his diet consisted mainly of foods that grew on the land. Rather than eating the greasy foods that she considered to be "killer soul," his meals included sorghum, beans, peas, peanuts, sweet potatoes, molasses, okra, wild greens, and corn. The most common sources of meat were eggs, wild-caught fish, and small game (such as opossum, squirrel, and rabbit), and these were usually cooked into a stew. Interestingly, pork was not plentiful and therefore was sliced into thin strips, preserved with salt, and used sparingly—never in a bubbling deep-fry. Flour was also expensive and therefore only used for baking on special occasions.

When Lynelle told him about my book project, he remarked that the African recipes reminded him of his childhood foods. In fact, the ndole sounded just like a type of gumbo stew that his mother used to make. Instead of bitter-leaf, they would use dandelion or collard greens. Instead of dried shrimp, they would use crab or crawfish. Lynelle had always thought that soul food meant ham hocks, chitlins, and fried foods and admitted that she was surprised by his comments.

My conversation with Lynelle prompted me to call Jessica Harris, PhD, a history professor, cookbook writer, and expert in African and African American culinary traditions. Harris's list of authentic soul foods very closely matched the list of childhood foods eaten by Lynelle's father. "You have to understand," she explained. "Most foods under the heading of soul food have their roots in slavery. The recipes traveled with the slaves from West Africa and then expanded from there. Of course, by necessity, these recipes incorporated the produce and spices as well as some cooking traditions of the white Southerners. For example, some meals were originally wrapped in banana leaves, but later this was changed to cabbage leaves.

"Given this history," Jessica Harris continued, "most people regard the original soul cooking as a diet of poverty. Sure they ate pork if they were lucky enough to get it. But, contrary to the 'high on the hog' diet of white plantation owners, their cuts of pork were quite 'low,' namely feet, snout, and innards." (Interestingly enough, these low parts are also lower in fat and calories and higher in nutrients!)

Jessica Harris believes that the transformation from healthy soul to "killer soul" happened in the past two generations. "The real problem," she said, "is when you take a diet of 'have not' and square it. Once people could afford it, they started eating too much fat and too much sugar. These days, people are eating super sized." Even in her childhood, Harris recalls, things were different. Dessert on most nights was just fruit, and on Sunday she would have a baked treat: homemade honey and nut Scripture cake. "Now," she laments, "the sweet tooth is out of control."

## Farmer Brown

Not too long after my discussion with Jessica Harris, I visited Jay Foster at Farmer Brown restaurant in San Francisco. Jay does not have opossum or squirrel on his menu, but otherwise he features foods that are pretty similar to those mentioned by Lynelle's father and Jessica Harris. Jay, a skilled chef and restaurant entrepreneur, is swiftly becoming a San Francisco food phenomenon. Recently he won the prestigious *Food & Wine* Magazine "Taste-maker" award given to people who have changed the world of food before age thirty-five. Farmer Brown, his current restaurant project, has brought new life to what was once a seedy corner of San Francisco's Tenderloin District. More importantly, it seems to be bringing new life to African American cuisine. At least, this is what I surmised after seeing the crowd of customers, many of whom are African American, that his restaurant attracts on a nightly basis.

Jay seems like an unlikely food revolutionary. Looking barely twenty, he was wearing Puma tennis shoes, sagging jeans, and a T-shirt bearing a picture of Richard Pryor. He has a slight build and almond-shaped eyes that he says come from his Thai mother. His taste buds, however, seem to have come from his Mississippi-born African American father. "My earliest cooking experiences were with my father's mother, Cora Glass. She cooked old-time Mississippi recipes. She had a farm, and that is where I learned to make things from the earth. Now I am trying to recreate her recipes with a modern twist. I am trying to give African Americans a healthy cuisine that represents their culture."

When I asked him what this cuisine was called, Jay shrugged. He did not seem to want to waste his time with definitions. "You can call it soul food, or neo-soul, or Southern, or just good cooking. What matters is that it tastes authentic." However, when I asked him what made his food so popular, he left no room for uncertainty: "It's the ingredients." Over 90 percent of the produce that Jay buys is organically farmed, and 100 percent of the meat is what he refers to as "sustainably raised." Most important to Jay, however, is the fact that most of the farms that sell him his food are small, local, and black-owned. Right now he buys mainly from three farms but has plans to buy from a dozen more in the future.

It was a Friday night, and Farmer Brown was hopping. I slipped into one of the booths in time to hear the woman in the booth next to me cooing over her plate of salmon croquettes and mustard greens. "Oh, these are fine!" I scanned the other tables where diners were enjoying dishes like roasted chicken, pulled pork, a stew that looked like gumbo, and another red stew that I guessed, after looking at the menu, must be jambalaya. While I did spot the occasional plate of french fries, what seemed to be consistently present on all the tables were plates of greens, yams, black-eyed peas, and green tomatoes. Aside from the food quality, I was struck by the modest sizes of the portions—not the heaping piles that one often associates with soul food.

I left Farmer Brown thinking that what I had just eaten defied any negative stereotypes about Southern or soul cooking. These were certainly not the killer soul foods that Lynelle and I had initially discussed. Instead, here was food that was moderately portioned, rich in vegetables, high in quality, not too salty, and certainly not too greasy. Yet the recipes were simple, and the cost to put them together (especially for San Francisco) was reasonable. It was not too much of a stretch to picture something similar being prepared by a Mississippi grandmother, like Jay's Grandma Cora Glass. In fact, many of the dishes reminded me of the recipes in Jessica Harris's cookbooks, cookbooks that, by her account, do not sell. Maybe times are a-changing, and the old soul is coming back. I called Lynelle again to tell her about my trip to Farmer Brown. She had already been there twice herself and was impressed. "I think those foods are similar to what my father used to eat. Maybe that is why he stayed healthy so long."

# Three Steps to Maintaining Colon Health the Cameroonian Way

## Step 1 (Basic):

- More *fiber*. Replace processed foods with less processed ones. Aim for a minimum of *25 grams of fiber* daily from foods such as fruits, vegetables, beans, nuts, and whole grains.
- Fewer unhealthy *fats*. Save deep-fried food for special occasions and avoid ordering deep-fried food in restaurants. Instead make slower-cooked stews using small amounts of unrefined oils (2 teaspoons per serving). Choose from olive oil and unrefined peanut, palm fruit, or sesame oil.
- Limit your intake of processed meats that are smoked or salted or that contain nitrates. At the BBQ, avoid overgrilling/charring animal proteins.
- Do some form of aerobic exercise for at least thirty minutes, four days a week.
- Cook at least one recipe per week from the Cameroon recipe section.

## Add Step 2 (Intermediate):

- More *fiber*. Substitute processed foods with less processed ones. Aim for a minimum of *30 grams of fiber* daily from foods such as fruits, vegetables, beans, nuts, and whole grains.
- Less *flesh*. Use meat as one flavor in a meal, not as a main dish.
- Eat more *fermented* foods. Choose from dairy products, pickles, brews, or other foods that are rich in beneficial bacteria.
- Increase your intake of fish, flax, and other omega-3 fats.
- Do some form of aerobic exercise for at least thirty minutes, five days a week.

- Cook at least two recipes per week from the Cameroon recipe section.

### Add Step 3 (Advanced):
- More *fiber*. Substitute processed foods with less processed ones. Aim for a minimum of *35 grams of fiber* daily from foods such as fruits, vegetables, beans, nuts, and whole grains.
- Include *foraged* greens with *folate*: Fill up on greens such as collard greens, mustard greens, dandelion greens, and spinach, that are rich in folate and other phytonutrients.
- Experiment with traditional African grains such as millet, sorghum, and teff, and try soaking the grains overnight to increase the bioavailability of the nutrients.
- Do some form of aerobic exercise for at least thirty minutes, six days a week
- Cook at least three recipes per week from the Cameroon recipe section.

# 9. Okinawa, Japan:
# A Cold Spot for Breast
# and Prostate Cancers

As odd as it sounds, all roads seemed to lead to the island of Okinawa. First I was contacted by the University of Hawaii and asked if I would be interested in a summer-long stint as a visiting medical professor at their collaborative teaching program on Okinawa. Then, several weeks later at a conference in Arizona, I met a Japanese emergency room physician named Kenji Hamada who happened to be from that very same island. Over breakfast with Kenji, I mentioned my interest in Okinawa, and he assured me that if I made my way there, he would be happy to give me an introductory tour. Finally, within that same time period, I met Connie and Winston, adult twins whose father was from originally from the island. They had come to see me with concerns related to breast and prostate cancer prevention, and it did not take much digging to discover that Okinawa was a remarkable cold spot for these two diseases. Next thing I knew, I was on a plane flying south from Tokyo, with the triple goal of investigating the guest teaching position, visiting Kenji, and learning what I could about Okinawan foods.

After skin cancers, breast and prostate cancers are the most common malignancies in women and men, respectively, and they are two of the top cancer killers. In many ways, the breast is for women what the prostate is for men. They are structurally similar, both being hormone-sensitive, fluid-secreting organs (otherwise known as glands) that develop at puberty when large amounts of hormones are surging through the body. In general, by the time a cancer in either of these glands becomes what we would call a true cancerous lesion, it has been growing slowly for ten to thirty years, and the cells involved in the cancer have gone through a multiphase process. There is good evidence that the right foods can help block the progression of these cancers in their early phases, thereby preventing the ultimate development of a full-blown (metastatic) cancer.

While Connie and Winston's Okinawan ancestors rarely suffered from breast and prostate cancer, they are part of a new generation of Asian Americans who are experiencing a dramatic increase in these diseases. We can all benefit by learning why their grandparents were able to remain healthy and cancer free. My journey to learn about indigenous Okinawan foods takes me to open-air markets and small home kitchens. I also rely on experts resources, both on the island and abroad, to help me understand the true nature of this remarkable cuisine. These experts include two Canadian brothers who are part of the Okinawan longevity research team and coauthors of a best-selling diet plan, an epidemiologist from Johns Hopkins, a biochemist from the University of Singapore, an emergency room physician from Naha City Hospital, a vegetable vendor, a psychology graduate student, and, last but not least, an Okinawan grandmother.

## Recent U.S. Breast and Prostate Cancer Statistics:

- One-tenth of women will get breast cancer, and one-sixth of men will get prostate cancer.
- Prostate cancer: 235,000 new cases each year and 27,000 deaths per year
- Breast cancer: 215,000 new cases each year and 41,000 deaths per year

## The Yoshikawa Twins

It is unusual enough for me to have a set of twins in their late forties in my medical practice. It is even more out of the ordinary for them to routinely schedule their appointments back to back. For this reason, my visits with the Yoshikawa twins made a lasting impression on me. Connie and Winston lived in their childhood home along with their parents and their ninety-seven-year-old grandmother. Connie was energetic, talkative, and very petite. At four feet eleven inches, her feet dangled off my waiting room couch, not quite reaching the floor. Winston was heavyset and a good head and a half taller than his twin sister. He would sit quietly next to her with a vaguely pained expression on his face. Clearly, Connie took her fifteen minutes of seniority over Winston very seriously as she treated him like a little brother: whenever she made an appointment for herself, she would also make one for him. When it came time for the actual clinic visit, Connie went first, and she would invariably slip in a suggestion about what I should do for Winston.

If you'll forgive the pun, the first time that I was treated to the Yoshikawa double-header, they were there to see me for a twin set of glandular issues. As part of an executive health program at work, Connie had done a screening mammogram. The result was abnormal, and she had just seen a surgeon for a diagnostic biopsy. Fortunately, the biopsy result did not turn out to be cancerous. However, it did show that there were some changes in her breast, called atypical hyperplasia, that were considered abnormal or premalignant. The breast specialist had suggested that Connie might want to start a medication to lessen her chance of developing breast cancer. While Connie was willing to consider this option, she was concerned about potential drug side effects and had made the appointment with me to discuss nonpharmaceutical approaches to breast cancer prevention—specifically changes that involved her diet and lifestyle.

Winston came in next. Blushing profusely and wringing his hands, he confided to me in a whisper that Connie had pushed him to make the appointment. He was there to talk about his prostate. His sixty-eight-year-old father had recently been diagnosed with cancer of the prostate gland and was scheduled to start radiation treatments the following week. Winston was not only worried about his father but also wanted to know what he could do to avoid the same fate. To make matters worse, he had recently noticed that his urine stream had weakened, and there was more dribbling afterward. He realized that these were symptoms typical of BPH (benign prostatic hyperplasia) and was wondering if this could be an early warning that he too was at risk for prostate cancer.

## Don't Forget Cancer Screening

While no cancer-screening test is foolproof, there are tests than can help detect breast cancer and prostate cancer at a relatively early phase. As a vital part of prevention, women should discuss mammogram screening and men should discuss PSA (prostate specific antigen) level testing with their health practitioner.

Connie felt that her abnormal breast biopsy "came totally out of the blue." Neither her Filipina mother nor Yuko, her paternal grandmother, who was originally from Okinawa, Japan, could think of anyone in their family who had been diagnosed with breast cancer or breast hyperplasia. Similarly, Hirose, the twins' father, who had recently been diagnosed with prostate cancer, had never bothered to get screened. This was because he did not know of any relatives with prostate problems. His father, who had recently died at the ripe old age of 102, always bragged that his urine stream was as strong as that of a teenage boy. He certainly never seemed to have problems with his prostate.

The hyperplasia found on Connie's biopsy was akin to the hyperplasia that was causing Winston's urinary symptoms. While breast or prostate hyperplasia is usually considered a benign health issue, the research shows that these atypical cells sometimes represent an early phase of a true cancer. Given the parallel nature of these two problems, I set out to locate one cold spot that could offer up valuable dietary information on both breast and prostate cancer prevention.

# Okinawa: A Breast and Prostate Cancer Cold Spot

Until recently, the subtropical islands south of Japan collectively known as the Ryūkyū Islands, or Okinawa, were best known for their military history. They were the site of a bloody battle during WWII and, to this day, they remain home to one of the largest U.S. military installations on foreign soil. In the past five years, however, Okinawa has developed an international reputation of a very different sort: the Okinawan lifestyle has become synonymous with healthy living and great longevity.

While this is not news to Okinawans themselves, it took the publication of two best-selling books, *The Okinawa Diet Plan* and *The Okinawa Program*, for the rest of the world to learn about their impressive life expectancy. These books, written by North American researchers Bradley Willcox, MD, and Craig Willcox, PhD, and Makoto Suzuki, MD, are based on the findings of the Okinawan Centenarian Study, a study that followed a group of already elderly Okinawans into their hundreth year (and beyond) to see what factors helped them make it disability free to this birthday.

After reading these books and the related research papers, I learned that Okinawans live an average of five years longer than Americans and are four to five times more likely to make it to age 100. However, there is something else that is equally astounding about the health of these very old Okinawans: they practically never get breast or prostate cancer. Their rates of these two diseases are less than one-sixth that seen among the same age group in the United States. Furthermore, if they do get one of these cancers, they have a 50 percent lower chance of dying from it. Connie and Winston's grandmother, Yuko, recently visited her native village in northern Okinawa and was surprised to learn that none of her contemporaries had ever received a screening mammogram. Apparently breast cancer was so rare in their generation that this screening test was not considered to be cost effective.

What could be the explanation for these low cancer rates on Okinawa? Could it simply be a matter of good genes? Certainly genetics are a consideration; however, the migration research shows that there are other issues to think about. Otherwise, why would Asian Americans like Connie,

Winston, and Hirose be experiencing more breast and prostate problems than their ancestors who grew up in Asia? This increase is so dramatic that within two generations, the death rates from these diseases have gone from one-sixth that of the general U.S. population to a rate that closely matches that of other Americans. Based on this dramatic shift it would seem, once again, that genes play a secondary role to other factors.

## Genes and Breast and Prostate Cancer

In breast and prostate cancer, genetic predisposition probably accounts for less than 10 percent of all cancer cases. While there are molecular tests for some of these genetic markers, it should be noted that these tests are hard to interpret. In general, the recommendation is to see a genetic counselor if

- two or more first-degree relatives (parents, siblings, or children) have the same cancer,
- a family member was diagnosed with cancer at an unusually young age (less than forty-five years old), and/or
- a family member carries a known genetic mutation.

But what are these other "factors"? As we have discussed again and again, this is a hard answer to tease out. There certainly are many things other than diet that may contribute to low rates of cancer and overall longevity on Okinawa. The authors of *The Okinawa Program* discuss how the elderly in their study exercise daily; live a low-stress, "island life" existence; and get a good deal of respect and support from their community. These elderly people also continue to stimulate their intellect through pursuits such as dance, art, business, and literature. Perhaps these activities and traditions fall away with a move to the United States, and this could contribute to poorer health. Furthermore, in the United States, Asian American women tend to have fewer children and have them later in life, both factors that are thought to increase the risk of breast cancer. However, when epidemiologists carefully analyze the migration study data, all of them seem to come to the same conclusion: While all these changes may play a role, the surge in breast and prostate cancer that is now occurring in the Asian American population can be explained by a radical change in diet.

## Spam Revered

*Change in diet indeed. When I asked Connie about her diet, she told me that she and Winston grew up on Spam. Or perhaps it would be more accurate to say they grew up eating Kentucky Fried Chicken, Top Ramen, Kool-Aid, white rice, and Spam. Their mother, Delphina, had moved to San Francisco from her town in the Philippines in her early twenties and immediately fell in love with U.S. convenience foods. These were foods that she had never tasted in the Philippines but that somehow reminded her of home. Spam and ramen noodles were vaguely reminiscent of the bean curd cakes and pansit noodles she ate in her childhood, yet they had the dubious advantage of more salt and a rapid preparation time. Kool-Aid reminded her of the fresh mangosteen and papaya juice that she drank each morning for breakfast, and Kentucky Fried Chicken and rice were an easy substitute for the grilled chicken that her mother would make each Sunday. Over the years, Delphina had slowly replaced any traditional ingredients with these more modern ones, until her daily meals bore practically no resemblance to what she had eaten as a child.*

*When Yuko, Delphina's mother and the twins' grandmother, emigrated from Okinawa in 1986 and moved into the lower floor of the Yoshikawa home in San Francisco, she was horrified to see what her family was eating. At first she attempted reform but was met with such resistance that she gave up. By the time that I met Connie and Winston (and also had the honor of visiting the whole family), ninety-seven-year-old Yuko was doing something that was unheard of in a close Asian family—she was cooking only for herself. At least twice a week, she would walk the six blocks to the Asian markets on Irving Street to buy her own ingredients. Meanwhile, the rest of the family continued to eat as they had for many years.*

## A Perfect Asian-Fusion Diet

Inspired by Yuko's vitality and the impressive breast and prostate cancer statistics from the Centenarian Study, I decided that Okinawa would be the perfect cold spot for Connie and Winston. After all, they did say they enjoyed Asian foods. Perhaps it was just a matter of getting them to eat more meals at home, to make some substitutions, and to give their grandmother's cooking a chance.

Before embarking on my trip, I assumed that Okinawan cuisine would be very similar to that of mainland Japan. I imagined sushi, terikyaki, sashimi—all the foods we get at home in our local Japanese restaurant. However, shortly after my plane took off from Tokyo, an Okinawan businessman sitting next to me set me straight. Through his meager English and my sign language, we managed to strike up a conversation. When I pointed to a picture of food in the in-flight magazine and said that I was interested in Okinawan food, he smiled and bowed forward several times in his seat: "Okinawan food very healing, happy food. But not so the same as regular Japanese," he said. Then he added as an afterthought, "Asian fusion."

I soon discovered that this gentleman was absolutely correct. Okinawan food is not at all the same as mainland Japanese food. It truly is "Asian fusion," borrowing from the healthiest parts of a wide range of cooking traditions. Looking at a map, this makes perfect sense: The islands of Okinawa are smack in the center of the East China Sea, closer to Taiwan and China than they are to the main island of Japan. Other Asian countries, including South Korea and the Philippines, are not that much farther away. For centuries, Okinawa has exchanged foodstuffs and recipes with these neighbors, either through friendly trade or forced colonization, and this has led to a truly eclectic cuisine. During my stay, I was treated to stir-fries and soups that were more reminiscent of Chinese, Thai, and Filipino food than standard Japanese. Even the ingredients seem to make their way from all parts of the continent: jasmine tea from China, a liquor called awamori made from Thai rice, stir-fries flavored with turmeric from Southeast Asia, and soba noodles in the Japanese tradition.

## The Makishi Market

I arrived in Okinawa in the early morning hours and caught a public bus directly to my bed-and-breakfast in the island's main city of Naha. There I received a warm welcome from the owner, Mrs. Nakahara. Besides running a bed-and-breakfast, Mrs. Nakahara has a side business of making awamori—the Okinawan rice wine delicacy. The walls of her tiny living room are lined with huge jugs of the liquor, each one flavored with a different tropical fruit. After offering me a hot shower and a sampling of some pineapple-flavored brew, she gave me an envelope from my friend Kenji. Inside was an itinerary that accounted for practically each hour of my time on the island. It was so detailed that it looked like a prepaid package vacation tour. I was touched by the effort but soon learned that this degree of kindness and hospitality was nothing out of the ordinary on Okinawa.

According to Kenji's note, I would be in the Naha central food market in less than twenty minutes. Sure enough, moments later, he appeared at the doorway and ushered me into a taxi waiting at the street corner. As the cab crept down Shuri Hill through the morning rush-hour traffic, Kenji explained that if I wanted to learn about food and the secrets of Okinawan longevity, my lesson needed to start at the Makishi Public Market. He had asked Shima Itokazu, a psychology graduate student, to be my personal guide and translator. Aside from being an avid cook, Shima had worked in the vegetable market in her twenties and knew her way around the place.

I have to admit that Okinawa, up until that point, did not appear to be the cold spot that I had anticipated. My doubts had set in early that morning when my plane touched down in Naha International Airport. As I entered the airport's reception lounge, I suddenly found myself surrounded by dozens of Japanese teenagers who appeared to be heading off on a school excursion. Despite the early hour, they sported elaborate gelled hairdos and were wearing carefully assembled "gangster hip" outfits. What really caught my eye, however, was that each one carried a paper bag emblazoned with a very familiar giant yellow M—the McDonald's breakfast meal. It did not get better on the drive from the airport to the bed-and-breakfast. The wide avenue that brought me through the city was lined with neon signs advertising fast food. Among them I recognized a number of KFC and A&W Root Beer signs and at least one golden archway. It all made me wonder whether sixty-plus years of U.S. military presence had made this island more American than Japanese.

Now, standing with Kenji at the main gateway of the Makishi Public Market, things did not look more promising. I found myself gazing at an elderly woman who was ladeling out greasy blobs from a deep-frying vat. A printed sign on the front of her stall read, "Okinawan Doughnut." None of this seemed in keeping with an island notable for its longevity and low cancer rates.

Despite this initial impression, I remained hopeful. After all, I had experienced a similar type of disappointment in my first hours at almost every cold spot. I just needed to explore further. Circumnavigating a throng of Japanese tourists and the doughnut lady, Kenji and I entered the main covered market. Suddenly we were in a maze of stalls bustling with more picture-snapping tourists as well as local Okinawans out doing their daily shopping. At first, all I saw were racks of blue jeans and other Western clothes, flip-flops, electronics, prebagged candy, bottled awamori, and piles of local souvenirs, including cell phone charms, and yes, cans of Spam! Finally, after what seemed like an eternity, we emerged into the bright sunshine on the other side. There, at a bus stop, Shima was waiting for us. She was a tiny woman with a beautiful smile and an incredibly expressive way of speaking. After making the introductions, Kenji rushed off to work and Shima motioned for me to follow her.

## Pay Dirt at the Farmers' Market

I tagged along behind Shima as we made our way through a series of crowded alleyways. All the while, I had absolutely no idea where we were headed. Finally, after what seemed like an eternity, we emerged into a clearing and came face to face with our first pay dirt. It was a stall selling yams, with deep red skins, their vibrant green leaves still attached. Behind the tubers stood the vendor, an energetic woman with pure white hair and a perfectly pressed apron. With great excitement, I realized that, besides Yuko, this was my first in-the-flesh encounter with a very old Okinawan. She looked so supremely healthy that I half suspected she had been sent by central casting to play the part of the Okinawan centenarian. (By the end of the week, I had met dozens of older folks who looked just as vital. As a matter of fact, I became a bit blasé and was only impressed when someone told me he or she was older than 100.)

I was dying to ask the vendor her age, but fearing that this would appear rude, I decided to ask if I could take a picture of her yams. She spoke very little English but seemed to appreciate my interest in her tubers. She came around the stall, picked up a yam, and said "*imo*" as she lifted it overhead. Then she ceremoniously broke it open to reveal a lustrous golden interior. Next she picked up a different root, and to my amazement, the drab brown peel was hiding an intense amethyst purple center. She explained to me partly with sign language and partly through Shima's valiant effort at translation that these tubers are yummy (smacking her lips and rubbing her belly) and good for your health and that they are one of the most popular foods in Okinawa. In the days that followed, I was served *imo* with meals, imo for snacks, and even had imo ice cream and custard for dessert. I also learned that after the Second World War, *imo* was literally life sustaining. It grew easily and quickly even in the postwar rubble and for several years was the main source of nutrients for many Okinawans.

Leaving the imo stand, we proceeded along the alley, and I quickly realized that we were in a giant produce market. At each stall, I would "ooh" and "aah" over the goods, while Shima engaged the elderly vendor in an animated discussion. After every stop, she would share some of her favorite ways to prepare that particular fruit or vegetable. Not far from the *imo*, we found piles of *goya*—dark green, spiky-skinned squash. Nearby were papayas, both ripe and unripe, some split open to reveal their fluorescent orange or light green flesh. On another table were neatly arranged bunches of greens with names like *handama*, *nigana*, and *yomogi*. They looked vaguely like dandelion or kale, but with a color so vibrant they almost appeared chartreuse in the sunlight. Some fruits and vegetables, like the tomatoes, leeks, and melons, were familiar, while others looked like very distant cousins to ones I know. As we neared the market's exit, I gazed back over the whole scene and realized that all this produce made up a fantastic rainbow of cancer-fighting antioxidants.

## Cancer Protection Direct from the Primordial Soup

But what is an antioxidant? The word is bandied about so often that, at this point, most of us are ashamed to admit that we have absolutely no idea what an antioxidant is or what it is doing inside our body. And why do certain fruits and veggies have such a strong antioxidant effect? Thinking back to the Yoshikawa twins, I wondered specifically about the link between these foods and the prevention of breast and prostate cancer.

Several weeks later, after I returned to San Francisco, I e-mailed my questions to University of Singapore professor Barry Halliwell, PhD. Being an authority on antioxidants in human health and disease, he was able to explain to me why that wonderful array of plant matter that I saw in the Makishi market could help prevent the growth of cancer cells. According to Halliwell, a good way to understand how antioxidants prevent the initiation of tiny cancer cells is to look back 2.5 billion years. This was a period long before our planet had any human life-forms, a time when blue-green algae covered the earth. To survive, these algae developed the ability to convert sunlight into energy. However, no biological process occurs without producing some waste or toxic by-product, and in this instance, the toxic byproduct was a substance called oxygen or $O_2$. For many of us it is surprising to learn that this gas we breathe in every three to five seconds can be a toxic "free radical," capable of causing cancerous DNA mutations as well as other biological damage.

As more oxygen filled the atmosphere, the surviving plants evolved special defenses in the form of *antioxidants* to deal effectively with this free-radical exposure (otherwise known as "oxidative stress"). Enter our ancestors, the early humans. When they first made their appearance as simple multicellular organisms, their survival also depended on their ability to tolerate oxygen and other toxic free radicals in their environment. They had two evolutionary options: they could start to make their own anti-oxidants, or instead of reinventing the wheel, they could feast on the surrounding plants that were already making these essential substances. Taking the easy way out, they did the latter. This explains why, from our earliest origin on this planet, we have depended on antioxidant-rich plants for our survival and why they continue to play a critical role in preventing or reversing chronic diseases, including cancer.

## Celebrity Antioxidants

When it comes to preventing breast and prostate cancer, are all fruits and vegetables created equal, or are there some that might offer extra protection? Search as I might, I could not find any research that conclusively linked *one specific* fruit, vegetable, or legume with cancer prevention. Nor did I find convincing evidence that any one supplement, derived from these nutrients, will ensure a life that is cancer free. *Overall, the greatest protection seems to be conferred on people who, over many years, eat a variety of fruits and vegetables, be they red, green, orange, purple, or yellow.*

That being said, there are a number of celebrity antioxidant-rich foods that consistently separate themselves from the crowd when it comes to preventing breast and prostate cancer. Not surprisingly, the Okinawan diet has more than its share of these notable free-radical fighters. First there are those fruits and vegetables that contain lycopene. Lycopene is a strong antioxidant in the carotenoid family. In men, it concentrates in the prostate gland, and studies have shown that men who eat lycopene-rich foods have anywhere from 25–80 percent reduction in the risk for prostate cancer. Some of this research even suggests that boosting lycopene intake from foods may help slow or stop the progression to metastatic cancer in those who already have a diagnosis of prostate cancer. The link between lycopene and breast cancer is not as strong, but there are still a number of studies suggesting that carotenoids in the diet (as opposed to supplements) play a role in protecting women from breast cancer and ensuring long-term survival in women who already have a breast cancer. While the greatest source of lycopene is cooked tomato products, this is not the only food that offers this nutrient. Within twenty square feet at the Naha market, I spied gua-vas, watermelon, and grapefruit, all lycopene-containing foods.

## Get Your Lycopene (15–20 milligrams per day)

| Food | Lycopene (milligrams) |
|------|------------------------|
| Pink grapefruit (1 medium) | 8.0 |
| Tomato (raw, ½ cup) | 8.3 |
| Guava (1 cup) | 8.5 |
| Watermelon (concentrated in the white area near the rind, 1 wedge) | 13.0 |
| Tomato sauce (½ cup) | 21.9 |

Source: USDA Nutrient Database

Note: Lycopene is fat-soluble. Therefore, cooking a lycopene-rich food (such as tomatoes) with a little bit of oil will increase your absorption of this nutrient.

Other uniquely powerful fighters of free-radical damage in the prostate and breast include foods that are rich in the glucosinolates (sulforaphane and indole-3-carbinol). These are found in greatest concentration in garlic and a whole host of cruciferous vegetables, including cabbage, cauliflower, Brussels sprouts, bok choy, mustard greens, kale, and broccoli. In a variety of experiments, glucosinolates and glucosinolate-containing foods have been shown to alter the metabolism of sex hormones and inhibit the development of hormone-sensitive cancers. Some researchers are so convinced about the nutritional importance of these foods that they are pushing for national dietary recommendation to suggest a minimum of *five servings of cruciferous vegetables* per week. An added benefit of these vegetables is that they are an excellent source of bioavailable calcium. Since Japanese don't traditionally consume dairy, these veggies (along with fish and tofu) satisfy most of their calcium needs.

## Give Your Glands Some EGCG

It is impossible to discuss the cancer-fighting effects of the Okinawan diet without mentioning tea. EGCG, a flavonoid in tea, is thought to be the most powerful antioxidant in the leaf. A wide variety of research studies have shown that EGCG helps to prevent or slow both breast and prostate cancer cells as well as other cancer types. While much of the research has been done on green tea, all nonherbal teas come from the *Camellia sinensis* plant and are likely to be beneficial. Most people I spoke with in Okinawa flavor their drinking water with green tea and drink at least three cups of straight green tea daily. Since the benefits of tea seem to be dose dependent, I say drink as much as you can stand.

## Getting the Most Out of Your Fruits and Veggies

Not only are Okinawans blessed with a diet rich in cancer-fighting fruits and vegetables, they also prepare them in the healthiest way possible. While many of us grew up boiling our vegetables or nuking them to smithereens, Okinawans give theirs a light steam or a quick stir-fry. Sure enough, the research shows that the best way to preserve the antioxidant activity in fruits and vegetables is to cook them with minimal water and for a short duration. In general, they should maintain their bright color and a bit of a crunch. The Okinawan recipes in this book give you cooking times that will achieve this effect.

In addition, it is important to remember that the colorful part of the fruit or vegetable is usually the richest source of antioxidants. This is often the peel or the skin, and therefore, you want to avoid peeling any produce where the peel is edible. Another good rule of thumb is to choose smaller fruits and vegetables, not only because they taste better but also because they have a higher skin-to-pulp ratio. For example, cherry tomatoes contain six times more of an antioxidant called quercetin than normal-sized tomatoes mainly because cherry tomatoes have a higher skin-to-volume ratio.

| Antioxidant-rich Okinawan fruits and vegetables that are linked to breast and prostate cancer prevention | Pro-oxidants foods that are known to cause oxidative stress |
| --- | --- |
| <ul><li>Garlic, onions, and leeks</li><li>Cruciferous vegetables: cauliflower, Brussels sprouts, bok choy, Chinese cabbage and regular cabbage, mustard greens, kale, and broccoli</li><li>Goya (bitter melon)</li><li>Tomatoes</li><li>Watermelon</li><li>Grapefruit</li><li>Guavas</li><li>Imo (yams)</li><li>Green tea</li></ul> | <ul><li>Heterocyclic amines (charred meats)</li><li>Foods contaminated with heavy metals, pesticides, or hormones</li><li>Saccharin</li><li>Select food dyes and colorings</li><li>Smoked foods and foods preserved with nitrates and phosphates</li><li>Trans fats</li><li>Satured animal fats</li><li>Alcohol (more than 1 serving per day)</li></ul> |

## Anticancer Foods from the Deep

As Shima and I were leaving the produce market, we wandered past what initially appeared to be a nursery specializing in shade plants. High above the stalls an opaque cloth canopy blocked out the bright midmorning sun. When my eyes finally adjusted to the lower light, I realized that what I took initially to be vines were actually long strands of seaweed draped over the tops of each stall. On trays, in pans and buckets, and tied up in bags were sea vegetables of all varieties. Some appeared as thick rubbery ribbons, while others were in small, matted clumps. In one metal bowl were sprigs of translucent pearl-shaped greens referred to as "sea grapes."

### Balanced Diet, Balanced Hormones

The high-iodine content of seaweed helps to increase thyroid function. This is an excellent counterbalance to some other Okinawan staples, including yams and soy, which can decrease thyroid function and are therefore considered goiterogenic. There is a direct connection between maintaining a healthy thyroid gland and your total glandular health (including mammary and prostate glands). Some scientists theorize that this offers yet another clue as to why we see so few cases of breast and prostate cancer in the Okinawan elderly.

Okinawans eat pounds and pounds of sea vegetables each year. Like maize in Mexico, potatoes in Iceland, and barley in Crete, this staple appears in an amazing array of traditional dishes: raw in salads, pickled as sauces and relishes, dried as a snack or a wrapper for rice and fish (nori), cooked in soups, and finally, sweetened in jelly or ice cream. In addition, seaweed is used widely in Okinawan herbal medicine to make traditional cosmetics and treatments for everything from asthma to hemorrhoids to stomach ulcers.

Besides being a low-calorie source of calcium, potassium, iron, iodine, magnesium, folate, vitamin K, beta-carotene, and fiber, seaweed has been shown to stop the promotion of cancer cells—especially in the mammary gland. What does this mean? It turns out that cancer initiation (or DNA damage) happens fairly often even in healthy bodies. However, in most cases, these abnormal cells undergo something called apoptosis, or cell death, and the cancer grows no further. Full-blown cancer only occurs when apoptosis does not take place and the abnormal cells start to rapidly divide and multiply. When researchers put these out-of-control cells in seaweed broths, they suddenly stop dividing and undergo apoptosis.

What is it in seaweed that can inhibit cancer growth? The research is just beginning, and no one knows for certain. Some scientists speculate that it could actually be the iodine since this is taken up preferentially by breast tissue much in the same way that lycopene is concentrated in the prostate. Others believe that the plethora of antioxidants (many yet to be determined) within the seaweed are the active ingredients. What is certain is that seaweed has a special place in the Okinawan longevity diet. When a research team studied the elderly in Yuko's native village of Ogimi, they found that those elderly who were the healthiest were also the ones who ate the most seaweed.

## Antiproliferative Shrooms

Mushrooms have practically no calories—but don't be fooled, these fungi are full of cancer-fighting substances that researchers are just beginning to understand. This is not news to traditional health practitioners throughout Asia who have been using mushrooms for medicinal purposes for thousands of years. Recently scientists have begun to identify some active ingredients in maitake, reishi, and shiitake mushrooms that stimulate immune cells to attack cancers. Clinical trials are now underway to see how extracts of these mushrooms can boost the effectiveness of chemotherapy and improve cancer survival. I believe that until more is known about extracts, it is probably best to get your mushroom nutrients the way Okinawans do—in soups and stir-fries. Mushrooms are also a good source of iron, potassium, zinc, copper, and folate.

## Dofu Heaven

Not far from the seaweed tent, we veered off into yet another market area. Shima explained to me that we were now entering the Makishi main market, the most famous market in Okinawa. Once again, I was greeted by dozens of stalls arranged in zones. To the right I saw the pickle people, their tables covered with a vast array of pickled food: ginger, seaweed, onions, green vegetables, and bowls of *mimiga*, or thinly sliced pickled pigs ears. Near the pickles, I found a series of vendors selling the essential ingredient in all Okinawan cooking: tofu, or *dofu* (as it is pronounced on the island). In display cases and on tables lay serene, white, perfectly rectangular blocks, some as large as a small suitcase. Sampling a piece, I realized that it had a consistency similar to our "extra firm" variety (an effect that is achieved by soaking the soy in a stiffening calcium solution).

Amidst the sea of white tofu, a small mound of a hot-pink–colored substance caught my eye. The vendor was too busy to indulge my questions, but seeing my fascination with the pink pile, she handed me a toothpick-full to sample. As a clutch of Okinawan shoppers watched me with amusement, I popped the unknown food into my mouth. A sharp, slightly bitter taste exploded on my tongue, and my nostrils were filled with an aroma reminiscent of rotten feet. Wanting to avoid looking like the hapless tourist that I was, I tried desperately not to gag. Suddenly I remembered my airplane companion's parting words: "Watch out for the pink food." This was *tofuyo*—tofu that is fermented and marinated in awamori rice liquor. During the week that followed, I encountered tofuyo on more than one occasion. Despite its alluring color and the knowledge that it was considered a great delicacy by many locals, I decided that my taste buds were not up to resampling this particular item.

## The Soy Dilemma

Standing amidst the rows of tofu, I faced the inevitable: To fully answer Connie and Winston's questions about breast and prostate cancer prevention, I needed to tackle the soy question. Are soy foods beneficial, harmful, or neutral when it comes to breast and prostate health? True, soy is an excellent source of digestible protein and is rich in folate, B vitamins, and calcium (if soaked in a calcium salt like the Okinawan tofu). But most of us have read or heard contradictory reports about soy and are left questioning the safety of this bean. Usually the reporters who are delivering these stories are getting their information from soy researchers who are also scratching their heads about the soy issue. In fact, there are few other areas of nutrition research that are as steeped in controversy and confusion.

This is what we do know for certain: Asians living in Asia eat a diet high in whole-soy foods, and they have the lowest rates of breast and prostate cancer worldwide. Unfortunately this is where the certainty ends. What role does soy play in preventing these cancers? Some researchers have focused their attention on weak hormone-like substances called isoflavones which they believe bind to breast and prostate cell receptors, thus blocking the effect of stronger hormones (estrogen and testosterone). Given the epidemiological evidence alongside this plausible scientific mechanism, their logical conclusion is that it's the isoflavones that play an active role in *preventing* breast and prostate cancers.

In truth, the answer is not so simple. Some studies do indeed show a benefit to boosting isoflavone levels, while others show no benefit, and some actually show that isoflavones might *promote* breast and prostate cancer. After my trip to Okinawa, I spent days culling through piles of research reports on the connection between soy and breast and prostate cancer. Then at a certain point, my head spinning, I put it all aside and called Bruce Trock, PhD, at the Johns Hopkins School of Medicine. He has spent a good part of his career studying the relationship between diet and breast and prostate cancer. Recently Trock published a summary (also called a meta-analysis) of eighteen different studies looking at soy and breast cancer, and he is currently involved in a long-term nutrition and prostate cancer study focused on men in Baltimore.

According to Trock, the isoflavone theory is all well and good, but it has one major flaw: There is no dose-response relationship. In other words, while his research and others do seem to show that soy foods offer breast and prostate cancer protection, what is confusing is that people with low blood-isoflavone levels often have the same degree of protection as those with higher blood levels. While there are a number of ways to interpret this data, Trock feels that the most likely explanation is that isoflavones, despite their seemingly obvious role in breast and prostate cancer prevention, may not be the only (or even the most critical) protective ingredient in soy. Given that we are not sure about which ingredients in soy are beneficial, he feels that the *kind* of soy food we eat may determine whether it offers protection or causes harm.

Retail sales of soy products in the United States now exceed four billion dollars per year, so there is little doubt that Americans these days are eating their soy. But what kind of soy are we eating? While Okinawans take in over 80 percent of their soy in a relatively unprocessed form as tofu, edamame, soy flour, soy milk, or miso, people in the United States eat a similar percentage of their soy in a processed form. Our soy foods are heated, mashed, and denatured to create a vast array of substances ranging from Tofurky to fillers for tuna fish to ice-cream sandwiches. Wretched-tasting soy isoflavone extracts also account for a high percentage of annual U.S. soy sales. These are doctored up with sugar and sold in the form of supplements, energy bars, and energy drinks.

These very different ways of eating your soy may have dramatically different effects on breast and prostate health. Bruce Trock believes that eating two or more servings of whole soy foods per day à la Okinawa is likely beneficial, whereas soy extracts and processed soy might be having a stimulatory effect on cancer cells. Some researchers even theorize that eating pure isoflavones may cause other problems related to excess estrogen influence, including hypothyroidism, infertility in men, and postmenopausal bleeding in women. Once again, while whole foods offer valuable protection, concentrated or denatured derivatives of these foods are having the opposite effect.

| Okinawan Soy Foods | American Soy Foods |
|---|---|
| *Anticancer Agents* | *Possible Cancer Promoters* |
| | |
| Tofu | Soy protein powders |
| Soy sauce | Isoflavone pills |
| Tamari | Energy bars/cereals with soy |
| Tempeh | extracts |
| Miso | Processed soy foods (texturized |
| Edamame | vegetable protein in fake hot |
| Natto | dogs, fake meats, etc.) |
| Unsweetened soy milk | |

## The Early Years Matter

Another twist in the soy story is provided by the Shanghai and Los Angeles adolescent soy studies. According to this research, it is not just the kind of soy that matters, but *when*, during your lifespan, it is eaten. Researchers in both cities have compared women who develop breast cancer with those who do not and have tried to understand what dietary and lifestyle factors make a difference. Consistently they have found that women without breast cancer ate more whole soy foods as teenagers! In one study, women who ate at least one serving per week of whole soy as teens had a 25 percent lower chance of developing breast cancer than those who did not eat soy in early life. Furthermore, women who ate soy as teens *and* as adults had a 50 percent lower risk. While this kind of research is not yet available for prostate cancer, there are some studies that suggest that dietary whole soy in early adulthood may play a similar role in men.

It makes sense that breast and prostate tissue would be the most vulnerable to mutation in adolescence and early adulthood, as this is the time when these glands do most of their growing and dividing. (This fact also explains an increase in breast cancer that was seen recently in Japanese women who were teenagers in the regions of Hiroshima and Nagasaki during the atom bomb explosions. While it took decades for their cancers to appear, it seems that they suffered damage to mammary tissue at a critical period in their development.) For those of us who may not have eaten much soy as teenagers, it is important to remember that it is never too late. Like many aspects of prevention, the longer you maintain a healthy habit, the better; therefore, better now than never. While whole-soy foods eaten only in later life might be less beneficial than soy foods throughout life, they certainly do offer some benefits.

## More Mojo: Start Low and Go Slow

*At her next visit, Connie raised another health issue. She told me that she had recently started to date someone. While she was really enjoying spending time with him, she was dismayed to see how uninterested she was in sex and wondered what had happened to her libido. In addition, she had started to have days each month when she would break into a deep sweat, especially at night. As a result, she felt like she was not sleeping very well and was often tired during the day.*

Listening to Connie, I was reminded of my conversation with Brad Willcox, a physician and one of the authors of the Okinawa diet books. He was impressed by the fact that the elderly people studied on Okinawa not only lived longer but also appeared much younger than their stated age. People were active well into their late nineties, and hardly anyone was confined to a wheelchair. (Hip fractures, which incapacitate many elderly in the United States, occurred 80 percent less often in the Okinawans.) Most of the Okinawan women reported easy menopause transitions free from hot flashes, sleep disturbances, and mood swings. In addition, the fatigue, poor memory, depression, loss of sexual drive, and impotence that we consider to be a normal part of aging were rarely experienced by even the eldest of Okinawans. In fact, the research team interviewed a number of men and women who were still experiencing healthy, active sex lives in their eighties and beyond. Willcox believes that the soy, the seaweed, and other traditional Okinawan foods are at least partly responsible for this phenomenon.

After doing lengthy laboratory analyses on the Okinawan elderly, the Centenarian Study research team was able to identify more specifically how these foods may be helping to preserve their vitality. The foods seemed to augment natural hormone levels since Okinawans had higher levels of thyroid hormone, cortisol, and sex hormones—including testosterone, DHT, estradiol, and dehydroepiandosterone—than a comparison group of elderly from the United States. Interestingly, if you look across the lifespan, Okinawans and North Americans seem to have different hormonal patterns. While the average North American starts out in adolescence with higher levels (a fact that many researchers attribute to the synthetic hormones commonly found in U.S. meat and dairy products), the hormone levels seem to fizzle out by the time most Americans reach their mid-fifties. Okinawans, however, tend to start low, increase slowly, and maintain their hormone levels longer than elderly in the United States. In fact, the researchers did an autopsy on one very elderly woman and were surprised to see that her adrenal glands (the organs where many of these hormones are produced) were the same weight as those of a much younger woman. Brad Willcox postulates that Okinawans may have a slower decline in hormones thanks to their excellent diet, which they continue to eat into their twilight years.

## Omega~3s for Your Glands

Moving away from the market's soy zone, I suddenly found myself surrounded by schools of the most brilliantly colored fish. Piled high on wooden tables, their fins moist and glistening, most were so fresh that they were not yet stiff. In front of the tables sat pails of giant green and yellow crabs, silver gray shrimp, and spiny sea urchins. From everywhere enormous fishy eyeballs stared at me accusingly. Here, flowing out of buckets and displayed on ice, was a treasure trove of omega-3s.

Okinawans, like Icelanders, get plenty of omega-3 fat—a fact that, no doubt, contributes to their impressively low rates of breast and prostate cancer. While there is some debate in the literature about whether *total* fat intake is linked to these cancers, there is little question, at this point, that *type* of fat makes a significant difference. Once again, getting your fats in an omega-3 form seems to be protective, whereas eating large quantities of omega-6-rich vegetable oils (such as corn, soy bean, or canola oil) or saturated animal fats may increase your risk of cancer.

Fish are also one of the few dietary sources of another nutrient that is critical for prevention of both breast and prostate cancer: vitamin D. Studies have shown that people who have higher blood levels of vitamin D tend to have lower rates of these cancers, and in the instances when a cancer does develop, it is generally slower growing and more treatable. You can get vitamin D in two ways: by exposing your face and arms to sunlight for at least ten minutes per day or from diet. Alas, both of these sources pose their own challenges, which explains why vitamin D deficiency is becoming more common. Many people who live in colder climates do not have enough access to sun year round (Icelanders, for example), while others who live in sunny places are slathering on the sunscreen to avoid skin cancer. In addition, vitamin D in a regular Western diet is hard to come by. Some dairy products and ready-made cereals are vitamin D fortified (whole-fat unfortified dairy also has small amounts), but this is usually not enough for your daily requirements. So what is a good source of vitamin D? At least two servings of fish per week!

## Where to Get Your Daily D

Recommended Daily Vitamin D Intake:

| | |
|---|---|
| 0–12 years | 400 IU |
| 13–50 years | 1,000–1,500 IU |
| 51+ years | 1,500–2,000 IU |

| Food | Vitamin D in International Units (IUs) |
|---|---|
| Cod liver oil (1 tablespoon) | 1,360 |
| Salmon (3 ounces) | 360 |
| Sardines (canned in oil, 3 ounces) | 231 |
| Fortified milk (1 cup) | 100 |
| Egg yolk (1 large) | 18 |
| Cheddar cheese (1 ounce) | 12 |
| Multivitamin containing cholecalciferol (vitamin D) | 400 |

Note: As little as ten to fifteen minutes outdoors (without sunscreen), three days a week should provide children and young adults with adequate vitamin D.

## Everything but the "Oink"

Deep in the heart of the market, I discovered one final food area: the pork section. Above the general din, I could hear the distinct sound of steel knives being sharpened. Men and women with muscles like prize fighters were slapping huge parts of pork onto the counter and slicing them up into every cut imaginable.

Like many other traditional Okinawan foods, pigs were introduced to the Ryūkyūs in 1392 by Chinese immigrants. The animals were able to thrive on the island thanks to the abundant supply of *imo* (sweet potatoes). Unlike mainland Japanese who, at that time, adhered to a strict Buddhist (vegetarian) diet, Okinawans happily integrated pork into their cuisine. This love of pork in all its forms has inspired local sayings like, "You can eat every part of a pig except for its oink."

Indeed, as I scanned the butcher counter, it seemed that only the "oink" was missing. There, in neat piles, were the usual ribs, chops, and shanks. But this was not all. Nearby were similarly large collections of feet, translucent yellow ears, furry snouts, rubbery skin, and coiled intestines. Propped high on a counter and presiding over the entire pig section was the severed head of a pig, its ears stuffed with fake flowers and a pair of dark sunglasses sitting on the bridge of its nose! As I started to snap pictures of this unusual display, several of the butchers came out in front of the counter and proudly posed for me next to their mascot.

I stood there in the hubbub thinking about Connie and Winston's love of Spam, and I wondered how all this pig could possibly be a part of an anticancer longevity diet. It seemed likely that, along with fast foods, Okinawan doughnuts and bags of candy, pork was yet another aberration in an otherwise wholesome cuisine.

As it turns out, I was wrong. My third day on the island I was fortunate enough to meet with Brad's brother Craig Willcox, coauthor of the Okinawa diet books and now a long-term resident on the island. Over a delicious Okinawan lunch served in the cafeteria of the nursing school where Craig teaches medical anthropology and gerontology, our conversation turned to the meat featured in our meal: pork. Craig believes that pork, as it is eaten traditionally on the island, is actually an important player in the longevity diet. He explained that the long-lived participants in the Okinawan Centenarian Study had high blood levels of proline and glycine, which came, at least partly, from the collagen and elastin in pork. These proteins help the body to build and regenerate normal tissue. Pork also happens to be an excellent source of selenium, an essential mineral that concentrates in the breast and prostate and acts as a building block for a powerful cancer-fighting enzyme called glutathione peroxidase.

## Get Your Selenium

(RDA for men and women is 55 micrograms selenium per day)

| Food | Selenium (micrograms) |
| --- | --- |
| Spinach (cooked, 1 cup) | 2.7 |
| Whole-grain bread (1 slice) | 11.2 |
| Sunflower seeds (¼ cup) | 25.4 |
| Anchovies (3 ounces) | 31 |
| Pork loin (4 ounces) | 37 |
| Halibut (3 ounces) | 39.8 |
| Brazil nut (1 nut) | 100 |

Source: USDA Nutrient Database

After my discussion with Craig, I realized that here, once again, was an example of how all meat is *not* created equal. Unlike Spam and other processed pork products eaten so often by Connie and Winston, the pig dishes eaten by the elderly Okinawans were perfectly in keeping with their concept of *nuchi gusui*, or medicine for life. There are three main factors that determine the quality of the pork: how it is raised, how it is prepared, and how it is served.

In Okinawa, most pork is raised on purple and orange *imo*. Certainly a pig eating such nutrient-rich food cannot help but be nutrient-rich itself. Spam, however, is from a pig raised on commercially grown grain with a fair amount of antibiotics thrown in for good measure. Preparation techniques also make a difference. In Okinawan dishes, the ribs and other pork parts are boiled for several hours to remove all the fat, and therefore, most meals feature only lean meat. Spam, by contrast, is made by grinding up a whole shoulder of pork and preserving the product with nitrates, salt, and sugar. Finally, unlike the classic Spam sandwich, pork never seems to be the main feature of a traditional Okinawan dish. Instead it is used to complement an array of lower-calorie traditional foods including onions, greens, gourds, tofu, peppers, mushrooms, fish, and *imo*. Soups, ranging from *leg tabichi* (seaweed, daikon radish, gourd melon, and pork) to the *nakami suimono* (pig intestines and mushroom soup) feature pork, but in most cases, it is barely perceptible among the surrounding vegetables and rich broth.

## Hara Hachi Bu

It was not until Shima and I parted ways and I ventured onto the second floor of the Makishi market that I understood the most important aspect of the Okinawan anticancer diet. Kenji had advised me that a trip to the Naha market was not complete without having a "carry-in." This entailed purchasing a fresh fish from the market and carrying it upstairs to one of the little cafeteria-style restaurants to have it prepared. After much deliberation, I finally selected an iridescent blue-green fish that was handed to me neatly wrapped in newspaper.

My package in hand, I tromped dutifully up the stairs and plopped my fish down on the counter of the first restaurant I came to. As luck would have it, I had selected the one restaurant that did not prepare "carry-in" food. My faux pas was clearly amusing as it elicited some giggles from the girls behind the counter. Recovering from my embarrassment, I found the right place and waited as the woman behind the counter deftly cleaned my fish. Minutes later she produced a heavenly smelling soup beautifully served in a modest-sized ceramic bowl. Floating in the soup's broth were small pieces of my fish along with ginger root, greens, daikon radish, and a clump of soba noodles.

Starving after my market odyssey, I felt a twinge of disappointment that the serving looked so small, and I wondered whether I should have purchased two fish. Then I peered sideways at the Okinawan gentleman seated next to me at the counter and realized that the soup that he was slurping was, if anything, smaller than mine. (By the way, in Japan it is considered acceptable—and even polite—to slurp, especially when eating soba.) So I sat down, resisting my temptation to gulp, and mimicked his slow, contemplative style, all the while trying my best to make my slurps as musical as his. To my surprise, by the bottom of my bowl, I felt just perfect. Not exactly full, but certainly no longer hungry.

What I had experienced was one of the essential rules of Okinawan eating: *Hara Hachi Bu*, or "Eat until you are eight parts full." When you talk to Okinawans, they all believe that this principle plays a critical role in ensuring their longevity. Indeed, scientific research seems to support this notion. *Hara Hachi Bu* or, in medical terms, calorie restriction, is consistently linked with low rates of cancer and heart disease and a longer, healthier life. It also helps explain why Okinawans maintain a stable weight between ages twenty and one hundred, while the average American puts on one pound per year after age thirty. Although the term *calorie restriction* sounds rather dire, it does not mean starving yourself to death. Instead, if you take the Okinawan example, it means eating mindfully (similar to the Cretan *Siga Siga*). This entails enjoying each bite or slurp and stopping before you feel completely full. For most people, calorie restriction translates into eating about 10–12 percent fewer calories than they would normally consume. Now, whenever I find myself mindlessly stuffing in food, I think back to that bowl of soup at the Makishi market, breathe in slowly, and try to practice my *Hara Hachi Bu*.

## The Motubu Marathon

The following day was a Saturday, and Kenji had, once again, planned my itinerary in great detail. At seven in the morning, he reappeared at Mrs. Nakahara's with Teieisan, his seventy-two-year-old father-in-law. Teieisan, the appointed driver for the day, turned out to be yet another Okinawan who looked at least twenty years younger than his age. After a typical Okinawan breakfast of rice cakes with seaweed and delicious green tea, we drove up island to Motubu to see the Okinawa aquarium. The aquarium, a massive concrete structure that sits on the coast, is considered to be one of the jewels of Okinawa. Indeed, the tropical fish, displays, and dolphin show rival any other world-class aquarium I have seen.

However, it was the goings-on in the streets surrounding the aquarium that really caught my attention. Kenji had warned me that Motubu would be extra crowded because there was a marathon taking place. Curious to see a Japanese-style marathon, I joined the spectators on the sidewalk and waited for what I assumed would be a cluster of elite runners. Sure enough, a handful of serious runners whizzed by. But then, to my amazement, chugging along the street came the most motley crew of athletes I have ever seen: children in school uniforms, middle-aged women with hot pink running suits, and dozens of extreme elders, some in running suits and others just wearing their street clothes. When I voiced my surprise to Kenji, he looked confused—apparently to an Okinawan, there is nothing odd about a herd of ninety-plus-year-olds jogging along in a marathon. . . . Aha! Another clue to this impressive Okinawan longevity.

## The *Okazu* Cook-off

My last evening in Okinawa I rejoined Kenji at his house for dinner. At the door, I was ceremoniously greeted by Kenji's wife, Chitose, and other members of the family, including Chitose's mother, Fujikosan. Fujikosan is a slim, energetic woman with a head of thick, jet black hair. Within minutes of meeting me, she made it her business to inform me that, despite her seventy-plus years, her hair was absolutely undyed.

While Teieisan and a gaggle of his grandchildren sat nearby watching sumo wrestling on television, Fujikosan and her daughters taught me how to prepare ten classic Okinawan *okazu*, a term used to describe any dish that accompanies the two essential parts of a meal: rice and miso soup. Within sixty minutes, we had made goya champuru; bean sprouts and green papaya in a sweet and sour sauce; bamboo and konbu salad; pork intestine soup (not my favorite); jushi rice (the favorite for the kids); eggplant with miso sauce; tofu and spring onion champuru; nigana aemono salad (mixed bitter veggies and tofu); deep-fried gurukun (the official Okinawan fish); and finally, *imo* custard for dessert.

Using Kenji as the translator, Fujikosan chatted to me about each new ingredient, occasionally pausing to give a quick directive to one of her daughters. The only incongruence in the whole experience was the can of Spam sitting on the counter amid the piles of chopped greens and other indigenous foods. I asked Fujikosan about it, and she pointed at it almost as if it were the first time she was seeing it. "Sa-pam," she said. Kenji explained that since the war, it has been considered a traditional Okinawan food and a delicacy. I suppose if it is a delicacy, by definition, that means that it is used sparingly. I recalled the Spam that I had seen for sale the previous day in a souvenir shop. At a steep 550 yen per twelve-ounce can, it deserved its place among the other precious trinkets.

After we finished cooking, I joined the family to pose for a group photo with the aromatic okazu in the foreground. As we knelt down around the low table to eat, I complemented Fujikosan on making such a feast. She laughed. Yes, ten okazu was a little more than usual but not much more. Even on a routine week night, she would not think of offering her family any less than five. As someone who rarely makes more than two dishes per meal, I thought this sounded amazingly ambitious. However, I had noted that Fujikosan managed to produce this buffet in about the same time that it might take me to make two larger portioned and slightly more complex dishes.

As we passed the okazu around the table, I noticed that Kenji's children and nieces enthusiastically served themselves scoops of the bitter goya champuru. How remarkable that this slightly acrid taste was appealing to their young taste buds! Certainly, my own school-age kids would not touch the stuff with a ten-foot pole. I watched as the adults put little servings of everything on their plate, sampling each in turn along with the occasional chopstickful of rice. Similar to the mezes on Crete, this concept of okazu was yet another health secret. Eating small amounts of an assortment of foods at one time offers a wider selection of nutrients and a greater opportunity for synergistic food reactions. Furthermore, at least in my experience, a greater variety of flavors helps me to slow down and prevents me from overeating. As we ate, Chitose complemented me on my chopstick dexterity, and I was thrilled.

## The Right Rice

In Japanese, the word for rice is also the word for *meal*; this is because rice is not an okazu, but rather an essential part of each meal. While the most popular rice eaten in modern day Japan is processed white rice, I spoke to a number of elder Okinawans who preferred *genmai* (whole-grain brown rice) because this was the rice they grew up eating. As compared to sweet white rice, genmai is a slow-release food with a nuttier taste, more fiber, more nutrients, and fewer calories; it also contains an antioxidant called tricin, which seems to help prevent the growth of breast cancer cells (at least in a test tube). The elder Okinawans' preference for genmai rice must certainly have contributed to their exceptional health.

## The Okinawan Paradox

On the morning of my departure from Okinawa, I entered a Lawson convenience store, part of a Japan-wide chain, to see what foods were being sold there. Unlike the food I had seen at the central market, every product on these shelves was hermetically sealed in plastic and Styrofoam, with labels that almost uniformly featured cartoons of a cute, smiling faces. Near the checkout, the Spam display announced the newest Spam product, "Spam Singles." I watched as two schoolgirls in matching uniforms bought a package of the singles, two bags of potato chips, and a lollipop with artificial tropical colors that rivaled those of the antioxidant foods in the market.

Next, I visited the Makiminato McDonald's, which is located on the west side of the island near one of the U.S. military bases. Opened in 1976, this particular restaurant has the distinction of being the second McDonald's franchise to open in Japan (the first was in Tokyo in 1971). Inside its air-conditioned interior, I saw middle-aged businessmen, elementary school children, teenagers, and young women. Unlike the trim vendors in the market, almost all the patrons and workers here were carrying extra weight. It was hard to believe that I was on the same planet as that open-air market, much less the same island. Here the only remnant of Asian culture was the iced oolong tea that McDonald's serves as a culturally sensitive alternative to the standard soft drinks—of course, there was also the McDonald's version of Chinese chicken salad.

I was witnessing first-hand what I had heard referred to as the "Okinawan paradox": the fact that this small island is home to the healthiest people in Japan alongside some of the least healthy. While some Okinawans are still eating their longevity food, others are consuming more hamburgers per capita than anywhere else in the country. As a result, although people of Connie and Winston's grandparents' generation are living past 100, younger Okinawans have lost their ranking as the longest lived of the Japanese. In fact they are now far behind most other Japanese prefectures. Obituaries are starting to report lists of people (mainly men) who have died of heart attacks, cancer, and strokes in their forties and fifties, something that was unheard of a generation ago.

Once again, as I saw the gleeful youngsters waiting to be served and watched the Big Macs and fries fly out the door of the Makiminato McDonald's, I was reminded of the irresistible nature of these modern foods. I thought of Connie and Winston eating their KFC and instant soups and drinking their Kool-Aid, while Yuko, standing inches away, prepared the same healing foods that I had just sampled in Fujikosan's kitchen. It is easy to see how the twins were lured in by the promise of sugar, salt, fat, and starch all in one big bite, even when the secrets of longevity were right within their grasp. Nonetheless, I was excited to tell them about my travels and reinforce their plans to start to cook like their grandmother.

## Eating Like an Okinawan

*The next time I saw Connie and Winston, I regaled them with stories of the Naha produce market. They both agreed that the fruits and vegetables, which made up about 70 percent of Yuko's diet, were largely absent from their own meals. They had not grown up eating many of them and had no idea how to prepare them. Winston decided that he was open to trying some of Yuko's foods, and Connie went so far as to say that she would accompany Yuko to the Irving Street market to see first-hand the full spectrum of fruits and vegetables (including sea vegetables) that were sold there. They also were making a concerted effort to increase the whole-soy foods in their diet and replace their standard white rice with either brown rice (genmai) or yams.*

*The last thing to change was their choice of meats. While it would not be truthful to say that they have eliminated Spam and other processed meats altogether, they are now limiting themselves to a small slice atop an otherwise healthy soup or salad. I thought about the can that I had seen on the counter in Fujikosan's kitchen and decided that this was probably the way that she enjoyed her "Sa-pam." After all, it seemed reasonable that one of the eight parts in Hara Hachi Bu could be reserved for special indulgences.*

## Three Steps to Preventing Breast and Prostate Cancer the Okinawan Way

### Step 1 (Basic):
- Practice the *Hara Hachi Bu* rule—eat only until you are about 80 percent full, or cut your total daily calories by 10–12 percent.

- Eat at least *four* servings per day of in-season vegetables (a serving is 1 cup raw or 1/2 cup cooked).
- Whole soy is best. Try to have two servings a day of unprocessed soy foods, including tofu, soy beans (edamame), tempeh, and miso. A serving is 4 ounces (1/4 block) of tofu or tempeh or 1 cup of cooked soy beans or unsweetened soy milk. Avoid manipulated soy products, including soy protein powders, soy isoflavone pills, soy-infused cereals, and energy bars.
- Get your fats and vitamin D from fish—two 3-ounce servings per week. (And get vitamin D levels tested if you don't get enough sun and you don't eat enough D-rich foods.)
- Eat pork (and other meats) in moderation. Prepare pork as a side dish or condiment, and add it to soups and stir frys. Boil it first or trim the fat to ensure a lean cut of meat. Whenever possible, purchase free-range meats.
- Do some form of aerobic exercise for at least thirty minutes, four days a week.
- Cook at least one recipe per week from the Okinawa recipe section.

### Add Step 2 (Intermediate)

- Eat at least *five* servings per day of in-season vegetables (a serving is 1 cup raw or 1/2 cup cooked). Make sure to have at least five servings of cruciferous vegetables per week. Cabbage, bok choy, broccoli, cauliflower, kale, and mustard greens are all great sources of sulforaphanes—anticancer agents in the body.
- Load up on lycopene by eating watermelon, guava, pink grapefruit, and tomatoes daily.
- Try brown rice or *Haiga* (white rice available in specialty markets that has not had the nutritious germ removed) instead of white rice. If you find this rice unappealing, try

mixing half white, half brown for a while until you get used to the stronger, nuttier taste of the brown rice.

- Replace your coffee habit with a green tea ritual. Enjoy at least 3 cups per day for antioxidant protection. Look for decaf varieties if you're caffeine sensitive.
- When you dine out on Asian food, order fish and vegetables and avoid fried foods, flour noodles, and white rice.
- Do some form of aerobic exercise for at least thirty minutes, five days a week.
- Cook at least two recipes per week from the Okinawa recipe section.

### Add Step 3 (Advanced)

- Eat at least six servings per day of in-season vegetables (a serving is 1 cup raw or ½ cup cooked).
- Start adding sea veggies into salads, soups, and stir-fries. Have fun trying lots of varieties—wakame, kombu, hijiki, arame, and dulse.
- Be creative with mushrooms. Eat them for their culinary and anticancer properties. Try shiitake, reishi, maitake, enoki, crimini, portobello, and more varieties in soups and stir-fries.
- Substitute in subtly sweet desserts. See if the *imo* custard or a bowl of fresh fruit can satisfy your sweet tooth.
- Try to cook at least three simple dishes per week from the Okinawa recipe section per week to improve your assortment of nutrients and promote food synergy.
- Move like an Okinawan. Exercise daily, take part in your community, and stimulate your mind as you age by continually learning new things.

# 10. Foraging for Indigenous Foods in a Modern World

## A Major Obstacle?

Recently I was asked to facilitate a nutrition workshop for health professionals at a diabetes conference in Monterey, California. Because of my interest in cold spots, our discussion quickly turned to indigenous diets. We went over some of the research and then started talking about the recipes and the foods that played an important role in chronic disease prevention. The room was packed, and the content of the workshop seemed to generate a lot of interest. At the end of the session, however, two doctors came up to me separately to voice their concerns about the subject. One thought that the ingredients might be too expensive and, therefore, inaccessible to her lower income patients, while the other thought the ingredients were too "ethnic" and would, therefore, be hard to find in your standard supermarket.

Both comments got me thinking. Is it realistic to think that people across this country could put this eating plan into action? I am personally able to find most of what I need each week by making my Saturday morning rounds to the local corner store, the farmers' market, Drewes Brother's butchers, and the nearby Mexi-tessen, all of which are no more than eight to ten blocks from my home. But I, and many of my patients, live in a neighborhood in San Francisco that has a reputation for being an epicenter for foodies, a true "gourmet ghetto." What is it like in the rest of the world? Could those of you who live in other places find these ingredients somewhere in your shopping sphere? I had just wandered the world to glean cooking secrets that could keep us healthy; now it was obvious that I needed to travel within my own country—or at least my own state—to find out how these cold spot secrets and recipes could become a part of our daily lives. I decided to go shopping.

## The Lesson of "Local"

Before setting off, I took the shopping list from Appendix A of this book and compiled a sublist of foods that I considered staple or "backbone" foods for most indigenous diets: dried legumes, grains, an array of fruits and vegetables, nuts, fish, eggs, chicken, tofu, dairy products, and of course, olive and peanut oil. I also made a list of specialty products that were particular to each cold spot region, including some select spices, seaweed, and more exotic vegetables and starches such as jicama, nopales, and plantains.

My first stop as I headed out of town was not too far from home but still beyond my usual shopping zone: the Allemany Open Air Farmers' Market. It was a Saturday morning, and the market was packed. Entering the central area of the market, I was struck by the amazing diversity of the crowd. This farmers' market happens to be at the intersection of several neighborhoods: to the east is Hunter's Point, which is home to the biggest African American population in San Francisco; to the south, the Outer Mission houses a predominantly Asian American community with a large representation of families originally from China, Laos, the Phillipines, Thailand, and Vietnam; and to the west is Bernal Hill, which is rapidly transitioning into a "yuppie" professional class neighborhood but still has many of the original San Franciscan families whose ancestral roots are in Italy and Portugal. Finally, directly to the north is the Mission with a large Latino population. All these cultures and their respective languages seemed to be at the market on this sunny morning.

Using my wicker basket as a battering ram, I entered the throng and made my way toward the row of vendors. It was mid-February, and root crops and greens were in season. Stopping at the first stand that I passed, I watched as an elderly woman filled her basket to overflowing with daikon, mustard greens, and bok choy all the while chatting to the vendor in Chinese. Next to her, a man dressed in Dockers sipped a latte while loading some bok choy into the top basket of his toddler's high-tech stroller. Some of the vendors had signs that said they were organic growers, others proclaimed that they were pesticide-free, and a few had no signs at all. All of them seemed to be doing a brisk business. I stopped at the egg stall, the cheese stall, the mushroom stall, and the olive oil stall. Then I checked out a couple of nearby trucks filled with live chickens and rabbits, freshly butchered fryers, and filleted fish.

Wandering back to the vegetables, I stopped at the Lou Vue Farm stall and admired the gorgeous array of greens. This is a family-owned farm near Fresno and one of the Vue boys told me that his parents have been selling at the Allemany farmers' market for years. While they do not use pesticides, their vegetables do not qualify as organic because they still use some commercial fertilizers. I picked up a bunch of kale and admired it. It was a brilliant purple-green, and the stalk was not too huge or woody; there was plenty of good tender leaf to cook on this one. For a total of three dollars, I popped two one-pound bunches into my bag. One stall over was an organic farm vendor also from the Fresno area. Their produce was equally as beautiful, but their kale was almost ten dollars for two pounds. Was it really worth the extra seven dollars to have *these* bunches of kale over the two that I had just bought?

It just so happens that the previous week I had asked this very same question of Brian Halweil, a senior researcher at World Watch Institute and the author of the book *Eat Here*. Brian has spent a good deal of time thinking about the costs and benefits of different types of food production. According to Brian, all things being equal, you would want to support the organic farmer because he or she is using farming methods that are probably slightly better for your health and certainly better for the planet in the long run. However, what is most important is that, whenever possible, you support food that has been produced locally.

Listening to Brian, I recalled Professor Harriet Kuhnlein's definition of an indigenous diet: "When a group of people use their traditional knowledge to make a complete diet using *local* foods." "Local" is a term we are hearing more and more when it comes to healthy eating, and there are local food movements sprouting up across the country in both rural and urban areas. I have heard food advocates use the phrase, "Make sure your food has a face." Like Brian, they believe that while it is important to know *how* your food is produced, it may be even more important to know *who* produced it. There are compelling political, environmental, economic, and health reasons to "stay local." In his book, Brian offers an alarming statistic: produce in the United States typically travels 2,400 miles from farm to plate. Just imagine how much gasoline you save and how much more you contribute to the local economy by eating locally. Perhaps the best reason of all, however, is that local food is fresher. By and large, fresher food has more nutrients, fewer preservatives, and tastes better.

Back to my kale: it might not be organic, but similar to its more expensive competitor it was local—certainly the seven-dollar difference between the organic and conventional bunch could add up for anyone on a limited grocery budget. Plus the small, tender leaves suggested that very little fertilizer had been used during my kale's lifespan. I decided to enjoy my conventionally grown local kale guilt free, but promised myself I would also try to shop organic whenever the price difference was not too huge.

I spent the rest of the hour wandering from stall to stall and buying ingredients that matched my list. En route I bumped in to a number of people I knew, and we exchanged pleasantries and tips on which stalls had the best produce. At the gates of the market, I paused and reviewed my shopping list. I still needed peanut oil, soy sauce, dried beans, and grains, but otherwise I had successfully found all the items on the list! In addition, I had socialized with some friends and managed to get a little exercise and sun in the bargain. Overall this was an extremely pleasant way to start a Saturday.

I headed north across the Bay Bridge toward Oakland. In downtown Oakland, a similar farmers' market was in full swing. It was equally diverse in terms of its attendees and its produce. I saw a vendor accepting a food stamp voucher from one of his customers. When I asked him about it, he told me that all the vendors at this market took food stamps. "After all," he said, "this is as good as money and everyone deserves good food." I later had a discussion with Paula Jones, who directs the San Francisco Food Systems Project, and learned that this situation, far from being unique, was actually a part of a state-wide initiative to have all farmers' markets in California accept food stamps.

## Cho Lon Moi Market: Sorting the Authentic from the Imposters

In downtown Oakland, I also made a stop at the Cho Lon Moi food market on Eighth Street. In open crates outside the market, I found an impressive assortment of fresh vegetables, including a variety of roots, greens, and exotic fruits. Unfortunately, without the vendors standing behind their wares, I was unable to figure out what was locally grown. Some of the vegetables looked exactly like the ones I had just seen at the farmer's market, while others looked larger-than-life. I wondered if these had been raised with a healthy dose of fertilizer and pesticide. I decided to avoid the vegetable-on-steroid look and pick ones that looked a little scrawnier but still appetizing. I figured that these, most likely, had been grown with fewer chemicals.

Inside the market I felt totally overwhelmed. There was an entire aisle dedicated to dried fish and another to soy sauce. I stood there clutching my shopping list, not knowing where to turn. Finally, I gathered up my courage and approached an older man stocking bowls near the front. He did not speak English but graciously led me to a younger woman in the back who was more than happy to help. I told her that I was looking for a good soy sauce and some seaweed. First she took me to the seaweed section, which had at least thirty varieties of dried and packaged sea vegetables. I started inspecting the packages, looking for a list of ingredients but quickly realized that many food items were not labeled. Of the packages that were labeled, some contained pure sea vegetables, while others also had a variety of preservatives and dyes. I cannot tell you what species of seaweed finally ended up in my basket, but I am fairly confident that it was preservative free. I figured it was only through taste-testing that I had any chance of becoming a seaweed connoisseur.

After locating the seaweed, the saleswoman led me over to the soy sauce aisle. I read bottle after bottle and was shocked at all the added chemicals. Thankfully, here too I was able to find a couple of choices that contained just water, soybeans, and salt. At the checkout, I paid four dollars for several pounds of vegetables, some seaweed, and a bottle of soy sauce. What a bargain! Leaving the store, I pondered the upsides and downsides of this experience. Here I had found an impressive variety at unbeatable prices. However, among the authentic indigenous foods lay a fair amount of chemically-treated imposters. I reminded myself that just because something is sold in an ethnic market does not mean that it is truly indigenous.

Leaving Oakland, I headed away from the greater San Francisco metropolitan area along the 580 expressway. The highway was lined with malls and new tract home developments. Pulling into towns and cities off the freeway, I came across two more farmers' markets that had fewer vendors and were certainly less bustling than the ones I had just visited but still seemed to be well attended. Near a neighborhood of modest single-family homes in Tracy, California, I pulled into the last open space in the parking lot of a Safeway food supermarket. Judging by the number of cars in the lot, it appeared that this was where most people in this town had decided to do their Saturday shopping. As I pushed my cart through the automatic doors, I wondered what my forage might yield.

## Finding a Safe Way to do Safeway

I bypassed the display of prepackaged cold cuts near the door and went directly to the cheese display. I was looking for a feta or any other simple sheep or goat cheese. The first one I picked up called itself a "Greek Wine Cheese." The cheese itself was off-white with swirls of magenta and orange. Flipping it over, I read the label. Along with a number of unpronounceable ingredients, it also contained gum and carmine blue 2 and red 40 food colors. I decided that at the very least these additives were unnecessary and at the very worst they were bad for me, so I put the cheese back. Nearby was a cheese called feta, but it contained partially hydrogenated corn oil, sugar, a mold inhibitor, and cornstarch in the ingredients. It too went back in the case alongside my first selection. Finally, after several tries, I found Mount Vikos feta: part-skim sheep milk, cheese cultures, and salt. It came all the way from Massachusetts, but its three basic ingredients made it the best candidate so far. I was about to drop the cheese in the basket, but suddenly I spied, hidden in the corner of the case, a more local brand that was comparable to the Mount Vikos in ingredients and cost only twenty cents more. I chose that one instead.

At the meat and fish department, the hunt resumed. I skipped over the meat on "clearance." Knowing that none of my recipes called for large amounts of meat, I decided that this was not a place to cut corners in my shopping budget. After all, lesser quality meat is generally much richer in unhealthy fats, antibiotics, preservatives, and hormones and poorer in protein, vitamins, minerals, and healthy fats. At the main counter, I found frozen salmon from Alaska that had been treated with food coloring for added pinkness, chicken from a farm somewhere in Indiana that was too plump and fatty to be free-range, and meat from a big stockyard in Iowa that was marbled with fat. Finally, I came across some local, fresh-caught sole and a frozen free-range chicken, both from a neighboring county. These I kept.

As I made my way around the perimeter of the market, I passed the vast cookie and cracker aisle. Pausing there, I picked up random boxes and was amazed by how many still contain partially hydrogenated fats in their ingredients, a fat known to cause cancer and heart disease. "Whole grain" was the latest front-of-the-box buzz word on many of the cracker products, but these words rarely appeared in the actual list of ingredients. I decided to pass on everything in this aisle.

Moving on to the dairy section, I was surprised by my options for local, organic, and/or free-range foods. I selected some Uncle Eddy's eggs (local and free-range) and a local plain yogurt without preservatives, sugar, coloring, or gum. I passed on the omega-3–enriched eggs as they were twice the price of Uncle Eddy's. I decided that I would get my omega-3s in other ways.

Gliding past a double aisle of sodas, I arrived at a giant end-of-aisle tortilla display. The choice was overwhelming. The first decision point was wheat versus corn. I knew that corn could offer an excellent nutritional compliment to beans, so I focused my attention on the corn tortillas. After reading half a dozen labels, the winner was obvious. While most brands had sodium benzoate as a preservative and some even had sugar or partially hydrogenated corn oil, La Tortilla Factory's ingredients were all straightforward: water, stone-ground organic corn masa, and lime. Plus, the tortillas were made a couple of counties over in Sonoma, California.

In the condiment aisle, ingredient lists became longer and my aging eyes began to feel the strain. I popped on my newly purchased reading glasses and got to work. Here I found two brands of low-sodium soy sauce without chemicals or caramel coloring and one type of rice vinegar made by the Nakano Company with only one ingredient: glutinous rice. In the oil section, I could choose from several brands of olive oil that were dated for quality (see the Crete chapter), and I also discovered some unrefined peanut oil. Out of curiosity I picked up a bottle of stir-fry sauce in the same aisle. The Okinawan champuru sauce in this book's recipe section has four basic ingredients, while this premade sauce contained over twenty ingredients, most of which were chemicals, sugar, and color. Comparing both sauces highlighted for me why it is worth our while, whenever possible, to make sauces and condiments from scratch.

On the next aisle, big bags of dried red and black beans and brown rice were being sold for roughly sixty cents per pound. These were not organic but nonetheless they seemed like excellent purchases. I compared them in quality and price to the canned beans and packaged instant rice across the aisle. The bulk options had more fiber, less salt, and fewer additives and were much, much cheaper. Further down the aisle I came across a display of Bob's Red Mill grains and selected a packet of ground flax seed and two others of millet and barley. In the bread section, I once again chose local and even found a brand that had whole grains in its ingredient list rather than just saying "whole grain" on the front of the package. I moved past the chip and liquor aisles without so much as a glance, and then, using every ounce of self-control, I also managed to pass the Krispy Kreme doughnuts display without stopping.

Finally, at the other end of the store, I emerged into the produce section. This was certainly the least crowded area of the whole place, and I was prepared for the worst. On display stands in the center of this section sat overly large, waxy, pale-colored fruits and vegetables, many of which were out of season for wintertime California. The apples (which happened to be in season) were mushy and had been shipped in from Chile. Bunches of kale with light-colored leaves as huge and tough as elephants' ears had originated in a big commercial farm near Fresno, California. Technically these were "local" but they did not look right: certainly not organic and probably raised with way too much fertilizer and pesticide.

To my surprise, I then discovered the "organics" section with healthy-looking in-season fruits and vegetables. Here the apples looked just perfect: smaller and a little less uniform but without mushy spots or a waxy coating. Near the organics, an "ethnic" section had many of the roots and greens that I had just seen in the farmers' market. I recognized fresh ginger, daikon radish, and a nice variety of squash. Reading the labels, most happened to be grown relatively near to Tracy. I spoke with one of the produce managers and asked him if this array of produce was typical of Safeways across the country. "We are trying to buy more local," he said as he rebalanced the kiwi into a perfect pyramid. I left Safeway feeling hopeful. Certainly farmers' markets or a backyard garden are your best options, but there is a way to forage for indigenous foods at a larger market like Safeway as long as you are selective and devote some time to reading package ingredients.

## From the Farm Stands of Elma, Iowa, to the Walled City of Wal-Mart

Retracing my path down the 580, I veered off on 880 toward Gilroy. My destination was the 219,000 square foot Wal-mart Supercenter, the biggest grocery store in central California. It was Dana Friedrich who made me realize that I needed to make this trip. Dana is a market team leader for Farm Fresh to You, an organic farm in Capay, California, that sells at farmers' markets and also makes homes deliveries throughout the San Francisco Bay Area. I was buying vegetables at Dana's stand the previous weekend and found myself admiring how skillfully she bunched her kale; it was obvious that she had spent a lot of time around vegetables. I started to talk veggies with her, and it turned out that, in addition to being a personal chef, Dana had grown up on a farm in Elma, Iowa, a small town in the northeastern part of the state. Ten years ago, much to the bemusement of town locals, her father had decided to convert his corn and vegetable crops to an all-organic growing system. He was inspired to do this after his own father died from a blood cancer that was believed to be caused by farming with pesticides and herbicides. Dana told me that now her parents enjoyed organic vegetables all summer and did a brisk business of selling them to others in the area. "How about the winter?" I asked. She winced. "Well, then the options aren't so good. They do dry and freeze some foods, but mainly there is Wal-Mart."

I looked up Elma, Iowa, on the Internet. Sure enough, there is a Wal-Mart supercenter about forty miles away. But then I noticed that there were a couple of markets that were closer to town than Wal-Mart; one was the Evenson Food Market in neighboring downtown Riceville. What did this market have to offer? I called and spoke with Bill Thompson, the market's manager. He seemed like a friendly enough guy, so I asked him about his produce section. I expected him to describe something with not much more selection than a Quick Mart or a 7-Eleven. His answer surprised me. "No, I get all kinds of local produce, and there also happens to be a farmers' market in town. The local Mennonite community farms without pesticides or nitrates, and they sell all their stuff there." When I asked him why he bought this local produce for his store, he did not mention a word about "organic" nor did he talk about the health or environmental benefits of not using pesticides. "I do it for two reasons," he said. "It tastes better, and it's cheaper."

It turns out that when Bill says *cheaper*, this includes the actual costs of the food but also the hidden costs. As he explained it, "In the grocery business, every day we get faxes and e-mails about processed and packages foods being contaminated by *E.coli*, salmonella, or this, that, and the other. We spend a lot of time pulling stuff off the shelves and throwing it away." Certainly grocers pay dearly for these recalls, as does anyone who eats the food that has slipped through the recall. I can see why Bill prefers fresh and local.

Looking again at the Elma Web site, I realized it was pretty warm for late February, but still the temperature at midday had barely reached thirty degrees Fahrenheit. I asked Bill what people ate this time of year. There was a brief silence on the other end of the line. "Well, some people grow or buy more than they need in the summer and can or freeze it for winter. But most folks end up driving the forty miles to Wal-Mart." Bill's own produce section shrinks way down and mainly features fruits and vegetables from overseas and he admits that even he finds himself at Wal-Mart now and then. It seemed that my foraging adventure was not complete without visiting the supercenter of all food supercenters—Wal-Mart.

At the Gilroy Supercenter, I entered a small walled city. There were at least forty cashiers, and I could not see from one end of the store to the other. It was a week after Valentine's Day and piles of on-sale, human-sized chocolate hearts and three-pound bags of Sweethearts lined the entry wall. As I moved through the food aisles, I could not believe the sheer numbers of canned and processed foods. Included in these were a vast selection of unhealthy, yet vaguely ethnic-sounding options: Danish ham in a can, Guerrero frozen burritos bearing the slogan "*Un Pedacito de Mexico*" (a small piece of Mexico), Chow Mein frozen meals, Filipino deep-fried lumpia, and something called Swedish rye, which was just white bread with brown food coloring. Most were loaded with vegetable oils, salt, sugar, refined flour, and preservatives. To my surprise, however, I did uncover some indigenous food gems hidden among the piles of junk. There was Rosarita Brand Mexican Queso Fresca that comes from Southern California, lime-treated corn meal for tortillas, some good oils and vinegars, bulk beans, and even a selection of spices and seaweed.

The produce section had piles and piles of shrink-wrapped, unhealthy looking fruits and vegetables that rivaled the center section at Safeway, but, here too, was a small selection of organic and/or locally grown offerings. I recalled what Brian Halweil had told me about Wal-Mart expanding their local food–buying program. "This did not come down from the head-quarters," he said "but from the local produce managers who are meeting the demands of their customers. In my area two Wal-Mart Supercenters are part of the 'Pride of New York,' which is a program that showcases New York State produce." I left Wal-Mart thinking about what a minefield of unhealthy food I had just wandered through. Yet, even here, it was possible with some work to forage and make out decently.

## Whole Foods, Not Always so Whole

On my way back up the 280 toward San Francisco, I took the Sand Hill Road exit and pulled into the parking lot of a Whole Foods Market in downtown Palo Alto. Squeezing my Honda in among the bank of Lexuses and BMWs, I was reminded that this particular area of Palo Alto has one the highest median household incomes in the world. Walking past a wonderful display of oranges from Florida "on special" at the entryway, I entered directly into the produce section of the market. Here everything is gorgeously arranged and freshly misted, making these foods look appetizing to even the most die-hard vegetable hater. Little artfully decorated signs stuck into each variety of produce described what it was, where it was from, and whether or not it was organic.

For comparison's sake, I made my way directly to the kale. It was organic and local but it was also five dollars for a half-pound. Moving along through the rest of the displays, I notice that organic to conventional is split about 60:40 and that there is a fair representation of local farmers. In general the prices at Whole Foods are almost 30–50 percent more than the farmers' market and maybe 25 percent more than Safeway. I walked past the fish and meat counters, which had a wonderful selection of fresh-caught fish and free-range, organic, or sustainably-farmed meats. These too were pricey, but given their health benefits, I felt they were worth the steep ticket—especially if one used them sparingly. Similarly, there was a range of organic milks, cheeses, and yogurts, many of which had a note on the label reassuring me that their cows spent a good portion of their time in real outdoor pastures.

As I approached the center of the store, however, I suddenly found myself in treacherous territory. Like most grocery stores, this center area is home to all the dry goods. Sure there were some bulk dried beans (two to four dollars a pound!), a variety of indigenous grains, some good oils, and excellent spices. But I also saw an overwhelming amount of food that was almost as processed and manipulated as anything I had examined earlier that day. True, they did not have as many chemical-sounding food additives such as sodium benzoate or sodium nitrate, but there was plenty of natural corn sweeteners, sodium, carrageen, guar gum, refined grains, soy protein isolates, and food coloring. I picked up a tofu "Giblet Gravy" that had over forty ingredients, many of which I could not pronounce. Another aisle was chock full of energy bars that uniformly had some type of sugar as their first ingredient. Patients often tell me that they use these as meal replacers, and I wondered if they were misled by finding them in a "natural" food store.

As I was checking out, I noticed that many fellow customers had their shopping carts filled to the brim with mainly processed and/or premade foods. While Whole Foods offered up a wide range of wonderful indigenous ingredients, it was definitely possible to do a less-than-100 percent healthy shop in this health-food store.

## Foraging Tips

- Shop at a farmers' market if there is one nearby.
- Consider starting your own backyard or community garden.
- Buy whole ingredients rather than things that have been premade. Avoid the packaged food sections of the stores.
- Try not to cut corners by buying low-quality meat products—these tend to be the worst for you.
- Whenever possible buy foods that are local and in season. Ask the produce manager to help direct you.
- Stock up on in-season foods and learn to can, dry, or freeze them for winter.
- Avoid foods that have ingredients you don't understand or cannot pronounce. If the ingredients say "enriched," it is likely that they are not whole foods.
- When in doubt, try to select foods with simpler packaging. These tend to be more local and have fewer additives.

# A Very Long Journey

Driving back up the 280 toward San Francisco, I felt like I was coming home from a very long journey—a thirty-six-month journey, to be exact. It all started with Angela, my trip to the Amazon, and my experience with the Jungle Effect. From there, I had slowly made my way around the world only to end up making a whirlwind tour up and down the northern California freeways. For a brief moment, I contemplated some of the more depressing images I had seen in my travels: the diabetic amputee in Creel, the thriving fast-food chains in Okinawa, and the concrete resorts and smoky hospitals on Crete. Then, of course, there was everything that I had just witnessed in my own backyard: how incredible that there are grocery stores the size of a small city with liquor, soft drink, and candy sections, each bigger than the produce department! I thought about how these same packaged goods were so desirable to people in developing parts of the world and how quickly they were replacing the healthy cold spot foods. Indeed, I saw the wrappers and bottles littering the trails, meadows, and beaches in even the most remote of cold spots.

But then, a few miles later, my mind turned to the positive, and I felt a spark of hope starting deep in my chest. I thought about indigenous food activists and advocates, including Nikki Rose and Kauila Clark, who were fighting to preserve or bring back disappearing food traditions; I thought about scientists like William Connor, Craig and Brad Willcox, and Antonia Trichopoulou, who were doing the hard work to establish the health benefits of these indigenous diets; I thought about Taurino, Madame, or Fujikosan, who were willing to pass on their wisdom and their precious recipes, and I thought about innovators like Jay Foster who were inspired to carry the torch—albeit with a *neo* twist. Finally, I thought about all the patients I had interviewed for this book and how bravely they had undertaken the most daunting task of them all: changing how they eat.

Suddenly my heart practically exploded with optimism as I charged up the freeway going too fast for my own good. If we can export these packaged modern foods with such efficiency, we certainly can export a renewed commitment to eating authentic, indigenous foods. I hope that the cold spot adventures and recipes in this book will offer a way to get started, a jumping-off point for understanding how local ingredients and food combinations can work best for each of us. Looking out at the bright green hills of the late California winter, I suddenly pictured hundreds—no, thousands—of little cold spots popping up around this country and, yes, around the globe.

Three

# Cold Spot Recipes

I cannot claim to have invented any of these recipes. In fact, no one can make this claim. Instead most are a synthesis of recipes given to me a number of times by different traditional chefs and home cooks. They, in turn, got them from their parents, aunts, uncles, and grandparents. I have tried my best to include my many sources in the acknowledgments and the cold spot chapters. Thanks to an amazing group of recipe testers, I feel that the final recipes will meet the approval of a wide range of taste buds while still remaining faithful to the traditional proportions and ingredients. In the rare instances where a recipe was given to me by one individual, I acknowledge this in the heading for that recipe.

## Indigenous Foods Scoring System

Each recipe is accompanied with an indigenous food scoring system to help you plan your meals. The score is for *one serving* of that particular dish In general. you want to aim for

7 to 9 points per *day* of
**Antioxidant ingredients**

2 points per *day* of
**Omega-3 ingredients**

3 to 4 points per *day* of
**Slow-release carbohydrate ingredients**

1 to 2 points per *day*
**Fermented ingredients**

## COPPER CANYON, MEXICO

ATOLE BREAKFAST DRINK                                    *Serves 4*

Atole, which is similar to pinole, is a nutritious morning drink that dates back to the Aztecs. I think of it as a drinkable breakfast cereal. It can be flavored with everything from honey to chiles. Traditionally, it is made with a pure corn masa base, although some versions include other grains for added taste and protein. I usually mix up enough of the dried ingredients to last me several weeks and then make it fresh each morning. You can also soak the grains in water overnight to shorten the cooking time.

*⅓ cup masa harina*

*⅓ cup baby oats (lightly blended until they are chopped but not a complete powder)*

*⅓ cup other grain such as spelt flakes (lightly blended)*

*3 cups water*

*½ teaspoon cinnamon or 1 cinnamon stick*

*1½ cups low-fat milk*

*1 tablespoon honey*

*½ teaspoon vanilla*

Dissolve the masa, oats, and other grain in the water in a saucepan. Add the cinnamon and bring to a boil, stirring constantly with a wooden spoon for about 15 minutes or until the mixture thickens.

Add the milk, honey, and vanilla and bring the mixture to a boil once again, stirring to prevent scalding. It should have the consistency of heavy cream. Additional milk may be added to thin the mixture, if desired.

Remove the cinnamon stick and serve hot in mugs.

VARIATION: For fruit atole: Add 1 cup pureed fresh fruit (strawberries, raspberries, peaches, guava, or mango), just as the atole is coming to its second boil, after adding the milk.

|  | Antioxidants Points | Omega-3 Points | Slow-Release Points | Fermentation Points |
|---|---|---|---|---|
| Native Food Score | 1 | 0 | 1 | 0 |

## BREAKFAST OR LUNCH BURRITOS

Mix and match the following four indigenous Tarahumara recipes for a variety of burrito meal options.

## TORTILLAS
*Makes 12–14 six-inch tortillas*

If you have ever had a freshly baked tortilla, then it is hard to go back to the packaged ones. I know that making your own tortillas seems daunting, but once you practice, it is surprisingly quick and easy. Unfortunately, homemade tortillas become stale within several hours after being baked on the griddle, so it is important to make a fresh batch for each meal. You can mix up several batches of the masa at once and store the extra dough in a sealed container in the refrigerator. It will keep for up to a week.

I usually buy my masa flour (*masa harina*) at a Mexican specialty store, although it is also available in many natural food stores and even in conventional supermarkets. See the Copper Canyon chapter for more details. If you would like to use a traditional tortilla press, you can buy one for under fifteen dollars in most Latin American specialty stores.

*Fresh ground masa or 2 cups masa harina*

*2 cups warm water*

Knead the masa and water with your hands until the dough is soft but not sticky. Important: Let the dough stand for 15–20 minutes before adding more water. If it is still too dry or too sticky, then add water or masa flour (as needed) to reach the desired consistency.

Using the palms of your hands, roll the dough into golf ball–size shapes and press into a flat disc. For a more perfect tortilla, place each ball of dough between two pieces of wax paper and press with a tortilla press or a rolling pin. Add flour as needed to keep the dough from sticking to the paper.

Heat an iron pan or nonstick skillet over medium-high heat and place the tortilla in the skillet. No oil is necessary. After 30–60 seconds, the bottom side of the tortilla should be brown. Flip and cook the other side another 30–60 seconds until the tortilla puffs slightly. Place the tortilla in a clean folded napkin to keep warm.

|  | Antioxidant Points | Omega-3 Points | Slow-Release Points | Fermentation Points |
|---|---|---|---|---|
| **Native Food Score** | 0 | 0 | 1 | 0 |

## TARAHUMARA EGGS WRAPPED IN A TORTILLA (HUEVOS MARIA CRUZ) *Serves 1*

This is a simple, flavorful breakfast that can easily be eaten on the run. For convenience and to limit the number of dirty dishes, I usually make my tortilla first, place it in a napkin to keep warm, and then cook up my egg in the same pan.

1 egg or egg white* (preferably free-range)

3 tablespoons finely chopped tomato

3 tablespoons finely chopped cilantro

1 tablespoon finely chopped red onion

Dash of salt and black pepper

1 freshly made corn tortilla

Sprinkle of queso fresco** or feta cheese

Salsa (see next recipe)

Beat the egg, then add the tomato, cilantro, red onion, salt, and pepper. Make a fresh tortilla (see above) or use a store-bought one. Scramble the egg mixture in a pan over medium heat, place on top of warmed tortilla, and top with queso fresco and salsa. Serve immediately.

* Nutrition Note: One to two eggs per day are a great source of protein and vitamin $B_{12}$. Research shows that this amount of egg does not increase one's risk for chronic disease including heart disease and diabetes. However, this may not be the case for people who are already diabetic. Therefore, diabetics should limit eggs to one to two breakfasts per week and, even then, try to have only egg whites.

** Queso fresco is a crumbly, white cow's milk cheese available at Mexican groceries. You can buy it packaged, but it tastes even fresher and more delicious if you can get a slab cut at the deli counter.

|  | Antioxidant Points with salsa | Omega-3 Points | Slow-Release Points | Fermentation Points |
|---|---|---|---|---|
| **Native Food Score** | 3 | 1 (if using free-range, omega-3–enriched egg) | 0.5 | 1 (cheese) |

## TARAHUMARA SALSA

*Makes about 1½ cups*

When it is tomato season, I cook up a weekly batch of this salsa. Maria Cruz taught me that the secret to making a rich-tasting salsa is to boil the peppers and tomatoes. You can mix this up easily in a blender, or for a chunkier salsa, you can make this by hand using a food masher or a mortar and pestle.

3 medium tomatoes or 5 tomatillos

3 medium jalapeño chiles with the seeds removed

4 large cloves garlic

Pinch of salt

1 cup loosely packed cilantro

1 tablespoon lemon juice

Bring a small pot of water to a boil and boil the tomatoes and jalapeños for 12 minutes, until skin is translucent. Drain and set aside.

Place the garlic in a blender and chop until fine. Add the salt, jalapeños, cilantro, and one tomato to the blender, and chop for only a few seconds more. *Do not puree.*

Add the rest of the tomatoes and chop for 1 second more. Stir in lemon juice. Store in a sealed jar in the refrigerator.

| | Antioxidant Points | Omega-3 Points | Slow-Release Points | Fermentation Points |
|---|---|---|---|---|
| **Native Food Score** | 2 | 0 | 0 | 0 |

## HOMECOOKED BEANS

Serves 8 to 10

Making a pot of beans once a week is a great habit to get into—the taste and nutritional value is far superior to canned, and they can be used in a variety of meals throughout the week. I choose a different bean each time for variety. Sometimes I eat them for breakfast with tortillas and scrambled eggs; other times I will blend them with a cup of broth and make a soup. For dinner I often add them to a stew or eat them over a grain (such as quinoa or rice), along with sliced avocado and some queso fresco. The epazote is a traditional Mexican herb that adds a nice flavor to the beans and, perhaps more importantly, helps cut down on bloating and gas.

If you make a big batch of beans, you can always freeze them. Freezing beans does not greatly alter their nutritional value. For more information on beans and alternate cooking methods, see Appendix B.

3 cups dried beans such as pinto, scarlet, black, anasazi, or white Northern

2 sprigs of epazote, fresh or dried (optional)

9 cups low-sodium chicken broth or water (add more liquid if you like soupier beans)

1 tablespoon olive oil or lard (in this instance, I often use lard because it adds a rich authentic flavor; see sidebar below)

1 cup chopped onion

5 cloves garlic, minced

1 teaspoon salt

Rinse beans, cover with water, and soak overnight or for at least 6 hours.

Discard water and rinse beans again. Place beans and epazote in a pot with 9 cups of low-sodium chicken broth or water and simmer for 1½ to 2 hours. Liquid should cover just about one inch above the beans. (I often use a simmer plate.)

While the beans are cooking, heat the oil or lard in a saucepan. Add the onions first and sauté until soft, then add the garlic and sauté for 5 minutes. Add the onion and garlic mixture to the pot of beans while they are cooking.

Important: Wait until the beans are soft before adding salty or acidic foods, or else the beans will not soften. Mash some of the beans to give a varied consistency.

Serve inside freshly made tortillas or over brown rice. Top with a sprinkling of queso blanco or feta cheese, green onion, avocado slices, chopped cilantro, and/or the Tarahumara salsa recipe above. Add a tossed green salad or jicama salad on the side.

VARIATION: Add 1 teaspoon ground cumin or 1 cup chopped tomatoes at the end of cooking for a different flavor.

| | Antioxidant Points | Omega-3 Points | Slow-Release Points | Fermentation Points |
|---|---|---|---|---|
| Native Food Score | 2 | 0 | 1 | 0 |

## SAUTÉED NOPAL CACTUS — Serves 4

Learning to deprickle nopales can be a bit challenging at first, but I have included some directions for those of you who are brave enough to tackle your own paddles. I must confess that I often buy my nopales prepeeled and even freshly chopped at the Mexican groceries. When shopping, look for nopales that are bright green, small, and tender, and avoid any that are flabby and soft. If you live in the Southwest or another desert region, you might be able to wander out your back door and pick some fresh paddles. Just wear gloves!

Besides being tasty, nopales have antidiabetic properties and are a good source of soluble fiber and antioxidants.

3 fresh nopal cactus paddles, spines removed (or if purchasing peeled and cut, about 2 cups chopped)*

1 tablespoon lard or olive oil

½ cup chopped white onion

3–4 ripe plum tomatoes, diced (1½ cups)

1 fresh serrano chile or ½ jalapeño chile, seeded if desired, minced

Salt and pepper to taste

⅓ cup chopped fresh cilantro

1 lime cut into wedges

*These can be found at Mexican grocery stores.

Clean paddles by cutting away the edge that outlines the paddle; this includes the blunt end where it was attached to the plant. Scrape off any stickers that remain and use the tip of your knife to dig out any that are embedded. Rinse under cold running water.

Stack up the cleaned paddles and slice them crosswise into ½-inch strips. Place the strips in a pot of cold water, cover and bring to a boil, lower heat and simmer for 10–15 minutes unit crisp-tender. Rinse well under cold running water and drain thoroughly.

Heat lard or oil in a large skillet over a medium heat. Add the onion and sauté until translucent, 2–5 minutes. Add cactus strips, tomatoes, and chiles. Stir until hot, 2–3 minutes. Season with salt and pepper, add the cilantro, toss, and serve with a wedge of lime. Use as a filling in warm corn tortillas.

VARIATION: Place sautéed nopales on top of vinaigrette-dressed baby greens and sprinkle with queso fresco or goat cheese.

| | Antioxidant Points | Omega-3 Points | Slow-Release Points | Fermentation Points |
|---|---|---|---|---|
| Native Food Score | 3 | 0 | 1 | 0 |

JICAMA-ORANGE SALAD                                       *Serves 4*

Jicama, also known as the Mexican potato, is native to Central America. Refreshing with a subtle, sweet taste, it can be eaten either raw or cooked. I like to slice it into strips to dip into guacamole or add to any green salad.

In this salad, jicama is mixed with other traditional ingredients— oranges, chili powder, and cilantro. Jicama is a great source of vitamin C, fiber, and slow-release carbohydrates.

### Selecting the Perfect Jicama

Jicama is available year round. Look for tubers that are firm and have dry roots. Make sure that the jicama has an unblemished skin.

3 cups orange or grapefruit sections, peeled (about 4 oranges or 2 grapefruits)

1 jicama, peeled (1½ pounds)

¼ cup fresh orange juice or lime juice

¼ teaspoon salt

2 tablespoons olive oil

1 dash chili powder

¼ cup chopped fresh cilantro

¼ cup toasted pumpkin seeds

Cut the citrus into bite-sized chunks. Peel the jicama and slice into strips that are about the size of a french fry—2-inch-long and ½-inch-wide strips. Combine the orange juice, salt, oil, and chili powder in a bowl, and then add the jicama strips and chopped citrus; toss gently. Cover; chill 30 minutes. Sprinkle with cilantro and pumpkin seeds.

| | Antioxidant Points | Omega-3 Points | Slow-Release Points | Fermentation Points |
|---|---|---|---|---|
| **Native Food Score** | 2 | 1 | 1 | 0 |

## MOUNTAIN STREAM FISH WITH CILANTRO AND VEGETABLES

*Serves 4*

This is a simple yet delicious way to cook a delicate white fish such as trout. In my home this has become a weekly standard. I just vary the vegetable according to what is in season. Here again, cilantro is used as a main vegetable rather than just as an herb. For those of you who like to fish and camp, you can make this dish cowboy-style over an open fire or a camp stove. Simply skip the step of warming the fish in the oven.

4 trout, boneless with head and tail removed (you can also substitute with other thin delicate fish such as perch or sole)

1 teaspoon salt

3 tablespoons olive oil (or substitute 2 of the tablespoons with lard for a smokier flavor)

3 cloves garlic, finely chopped

4–5 cups any green vegetable (use
in-season vegetables, including
asparagus, green beans, squash cut
in strips, or snap peas)

½ cup fish or vegetable broth

½ jalapeño pepper, seeds removed,
diced

3 tablespoons lime or lemon juice

⅔ cup chopped cilantro

Preheat the oven to 150°F. Pat the fish with salt. Heat oil in a skillet over medium-high heat and cook the fish on each side for 2–3 minutes, until the fish is toasty brown. Transfer fish to a baking dish and place in the oven to keep warm.

Turn the heat down and cook the garlic and vegetables for 1–2 minutes. Add the broth and jalapeño and turn the heat up for 2–3 minutes to intensify the flavor. Add the lime and cilantro, stir gently, and then pour over the fish.

| | Antioxidant Points | Omega-3 Points | Slow-Release Points | Fermentation Points |
|---|---|---|---|---|
| **Native Food Score** | 3 | 2 | 0 | 0 |

## THREE SISTERS STEW

Serves 4 to 6

This is a hearty, satisfying, easy-to-assemble stew that makes terrific use of a variety of slow carbohydrates. I tasted half a dozen variations on this stew during my time in Copper Canyon.

If you decide to use homemade pozole corn and home-cooked beans, then these will need to be prepared ahead of time.

3 slices pork or turkey bacon
chopped in bite size pieces or 1
tablespoon lard (optional but adds a
lot of flavor)

2 tablespoons olive oil

1 onion, chopped

2 cloves garlic, chopped

1 bell pepper, chopped

½ to 1 teaspoon dried ancho chile (or
more if desired)

2 teaspoons finely chopped fresh thyme, or
1 teaspoon dried

2 cups lima or other broad beans, soaked and cooked

2 cups fresh or frozen corn or 2 cups precooked pozole or 2 cups canned hominy

2 cups sliced delicata squash (the long, thin yellow ones with the green lines)*

3 fresh tomatoes, diced, or one 14-ounce can diced, tomatoes or 5 small tomatillos, hulled and diced

1 cup chicken or vegetable broth or water

1 bunch cilantro, chopped

In a large sauté pan, cook bacon over medium heat. Add the olive oil. If using lard, heat lard and olive oil at the same time.

Add the onion, garlic, and bell pepper and sauté until the onion is translucent. Add the chile and thyme and stir several times. Add the beans, pozole, squash, tomato or tomatillo, and broth and bring to a bubbling boil then turn down to a simmer and cover. Do not remove the lid! Cook for 20–25 minutes. For a thicker stew, remove 1 cup and puree in the blender, then return it to the general pot. Add salt and pepper to taste.

Add the cilantro and stir for 1 minute.

Serve warm topped with mashed avocado and a sprinkling of queso blanco and toasted pumpkin seeds. This stew is also delicious the next day with tortillas or eggs.

*To prepare delicata squash, peel with a vegetable peeler, cut off ends, then cut in half lengthwise and scrape out the seeds. Cut each piece in half again lengthwise, then crosswise into thin ½-inch-wide slices. These can also be substituted with a variety of summer squashes or other gourds, depending on what is in season.

| | Antioxidants Points | Omega-3 Points | Slow-Release Points | Fermentation Points |
|---|---|---|---|---|
| Native Food Score | 3 | 0 | 1 | 0 |

## Preparing Pozole

Pozole is a meaty corn grain. In general, 1 cup of dried pozole nets 2 cups cooked.

Option 1:
Soak the pozole overnight in enough water to cover it by 2 inches. Drain. Place in a saucepan with fresh water to cover by at least 1 inch. Bring to a boil and then simmer, partially covered, until the corn kernels are tender, about 2–3 hours. Many of the kernels will split open.

Option 2:
Prepare pozole in a slow cooker. Add a ratio of 1 cup pozole to 3 cups water. Cook for 4 hours at the low setting.

## FEAST: CHICKEN POZOLE SOUP

Serves 4

This warm pozole soup with juicy chunks of chicken and chile spice is perfect for a cold, winter day. Traditionally, pozole is served for holiday feasts. Be sure to add the garnish; it is essential to this dish.

1 whole free-range fryer chicken, skin removed, cut into serving pieces

1 small onion, diced

5 cloves garlic, minced

4 dried whole guajillo chiles (long, red dried peppers found at Mexican groceries)

2 tomatoes, cut into chunks, or one 14-ounce can chopped tomatoes

One 29-ounce can of hominy (or make your pozole from dried kernels as described in the sidebar)

Salt and pepper to taste

1 cup shredded cabbage

2 radishes, sliced thinly

½ cup crumbled queso fresco

Put the chicken, onion, and garlic in a pot and add enough water to cover. Bring to a boil. Turn down the heat and simmer for about 30 minutes until the chicken is cooked.

While the chicken is cooking, boil the chiles in a separate pot of water until soft, about 10 minutes. Let cool. Drain, cut in half, and remove seeds.

Blend chiles and tomato chunks in a blender or a mortar and pestle. You may choose to push puree through a sieve if blenderizing does not break down tomato and pepper skins. Add puree and hominy (or pozole) to the chicken pot.

Simmer for another 15 minutes. Remove chicken breast, chop into chunks, or shred and add back to soup. Season with salt and pepper. Serve warm in bowls. Top with shredded cabbage, sliced radish, and crumbled queso fresco. (Note: this stew tastes even better the next day.)

| | Antioxidant Points | Omega-3 Points | Slow-Release Points | Fermentation Points |
|---|---|---|---|---|
| **Native Food Score** | 3 | 0 | 1 | 1 (cheese) |

## ABOUT DESSERTS

Traditionally, fruit is the main dessert eaten in Copper Canyon. One of the most memorable treats, I had there was a whole mango dusted in hot chili powder. The deep orange and red were gorgeous together, and the two tastes melded perfectly. The mango was peeled whole, and then the flesh was sliced on a diagonal so that it looked like a rose. The fruit was then stuck on a sharp wooden stick, rolled in chili, and served like a Popsicle.

## ORANGES WITH CINNAMON                                        *Serves 4*

This dessert is fit for a feast, and yet it is surprisingly simple and healthy. It contains cinnamon, a spice which is known to have antidiabetic properties.

4 oranges in season, chilled

1 tablespoon honey

½ teaspoon cinnamon

Peel oranges and cut into ½-inch slices crosswise so that each piece is a perfect circle. Lay them out on a serving platter.

Sprinkle with cinnamon and drizzle with honey. Chill before serving.

|  | Antioxidant Points | Omega-3 Points | Slow-Release Points | Fermentation Points |
|---|---|---|---|---|
| Native Food Score | 2 | 0 | 0 | 0 |

## CRETE, GREECE

Many morning meals on Crete could easily pass for a lunch in other parts of the world. I love the idea of starting the day with a dose of fresh vegetables and a little bit of garlic thrown in for good measure.

## HORTA OMELET OR SCRAMBLE                                        *Serves 2*

This is a breakfast that can easily double as a lunch or dinner. The secret to making a delicious omelet is to use high-quality, free-range eggs. Squeezing fresh lemon over your omelet enhances the flavor and increases your absorption of the nutrients in the greens. I often serve this dish with tomato slices (when they are in season) and whole-grain toast.

2 tablespoons extra virgin olive oil

2 garlic cloves, minced

2 cups chopped fresh greens (such as purslane, kale, Swiss chard, spinach, carrot greens, beet greens, or a mix—remove center woody stems before cooking)

2 tablespoons crumbled feta or other slightly salty sheep or goat cheese

3 or 4 eggs, lightly beaten

3 tablespoons, kalamata olives, chopped (optional)

Salt to taste

1 lemon, cut into wedges (optional)

Heat olive oil over medium heat, add garlic, and stir until soft but not too brown. Add greens and stir until soft.

Evenly distribute your greens on the bottom of the skillet and then sprinkle with the feta.

Pour eggs over the top and cook until eggs are just as you like them. (You can cover with a lid to hasten the cooking time.)

Top with olives and sprinkle with a tiny bit of salt. Serve with lemon wedges.

|  | Antioxidants Points | Omega-3 Points | Slow-Release Points | Fermentation Points |
|---|---|---|---|---|
| **Native Food Score** | 2 | 1 (if using free-range, omega-3–enriched egg) | 0 | 1 (cheese) |

## A LITTLE BIT OF WINE IN THE MORNING NEVER HURT ANYONE

*Serves 2*

This is perfect for breakfast or lunch. My friend Eleni tells me that her 100-year-old Greek mother has made this her standard breakfast for the past 80 years!

*3 tablespoons good-quality olive oil*  
*1 tablespoon red wine*  

*Salt*  
*Bread or barley rusk for dipping*

Pour olive oil in a saucer. The quality of the oil really matters here. Add red wine to the center. Drizzle with a pinch of salt. Dip bread or barley rusk in the olive oil/wine mixture and enjoy.

VARIATION: Scoop on some mashed lentils and chopped tomatoes.

| | Antioxidant Points | Omega-3 Points | Slow-Release Points | Fermentation Points |
|---|---|---|---|---|
| **Native Food Score** | 1 | 0 | 1 (if served on whole-grain bread) | 1 (wine) |

. . . . . . . . . . . . . . . . . . . . . . . . . . . . . . . . . . . . . . . . . . . . . . . . . . . . . . . . . . . . . . . . . . . . . . . . . . .

## STELIOS'S BARLEY RUSK ROLLS *Makes 8 to 10 rolls*

. . . . . . . . . . . . . . . . . . . . . . . . . . . . . . . . . . . . . . . . . . . . . . . . . . . . . . . . . . . . . . . . . . . . . . . . . . .

The traditional Cretan bread or rusk is a special type of hard bread that is called *paximadi* on Crete. Although this recipe takes some time and practice, I encourage you to give it a try. Making your own bread certainly has its advantages. First of all, you burn a good deal of calories in the mixing and kneading process. (I always find myself in a deep sweat.) Also, by using the traditional olive oil and whole grain ingredients, you end up with a slow-release, nutrient-filled bread.

3⅓ teaspoons yeast (about 1½ yeast packets)

1½ cups luke warm water

2¾ cup whole-wheat flour

2¾ cup barley flour, including the bran

1½ teaspoons salt

½ cup olive oil

5 teaspoons red wine

In a mixing bowl, dissolve the yeast in all the water and stir in half of the whole-wheat flour to make a thick batter. Let this rise about 10 minutes.

Place the barley in another bowl and make a well in the middle. Add the salt, olive oil, wine, and yeast mixture. Stir until shaggy. Turn out onto the counter and start kneading, adding in the remainder of the whole-wheat flour as needed to make a thick, smooth dough.

Note on kneading the dough: Place on a clean, floured surface. Pick up the far side of the dough and fold it toward you, then push into the dough with the heel of your hand. Slowly turn the dough in a clockwise direction, repeating this movement. Get into a good rhythm. The goal is to knead the dough until you have formed a thick, smooth mixture that does not stick to the counter. This should take about 10 minutes, and your arms and stomach muscles should feel like they have had a little workout.

Place the dough in a bowl coated with olive oil. Cover with a clean towel and place in a warm place such as a sunny counter to rise for about 1 hour. (You can also cover the dough with a folded blanket to keep it warm.)

Place the dough back on a floured surface and knead for 5–7 minutes. Form it into rolls that are a little smaller than standard hamburger buns. Using a sharp knife, score the top of each roll a couple of times. Place them on an oiled baking sheet and re-cover with the towel. Allow to rise another hour. Bake at 350°F for 1 hour.

Remove from the oven; cool and cut in half horizontally. Return these to the oven and continue baking at 150°F for another 5 hours.

The hardened rusks will keep for weeks in a dry, cool place. Use in the following tomato and feta *dakos* sandwich recipe, spread with honey, or dip in milk or coffee.

|  | Antioxidant Points | Omega-3 Points | Slow-Release Points | Fermentation Points |
|---|---|---|---|---|
| **Native Food Score per rusk** | 1 | 0 | 1 | 0 |

## DAKOS—TOMATOES AND FETA TO-GO SANDWICH

*Serves 1*

This classic Cretan creation is a cross between a sandwich and a pizza. If you are using a real homemade barley rusk, then you need to soften it by dipping it for 5 seconds in a bowl of warm water.

*2 thick slices tomato or ⅓ cup roughly chopped tomato*

*2 slices whole-grain bread, toasted or 1 barley rusk (see previous recipe)*

*1 teaspoon chopped fresh oregano, basil, or parsley*

*1 teaspoon olive oil*

*1 tablespoon feta or goat cheese*

Assemble sandwich by placing tomato on top of bread slices, add herbs, drizzle with olive oil, and sprinkle with cheese.

|  | Antioxidant Points | Omega-3 Points | Slow-Release Points | Fermentation Points |
|---|---|---|---|---|
| **Native Food Score** | 2 | 0 | 1 | 1 (cheese) |

* * * * * * * * * * * * * * * * * * * * * * * * * * * * * * * * * * * * * * * * * * * * *

## WILD GREEN SALAD          *Makes 4 appetizer portions*

* * * * * * * * * * * * * * * * * * * * * * * * * * * * * * * * * * * * * * * * * * * * *

I make this salad at least four to five nights a week to accompany other dishes. You can vary the flavor by adding different fresh herbs.

*2 tablespoons extra virgin olive oil*

*2 tablespoons balsamic or white wine vinegar*

*1 clove garlic, minced*

*1 teaspoon lemon juice*

*1 teaspoon chopped fresh oregano, marjoram, thyme, or mint*

*6 loosely packed cups wild greens of your choice (such as chicory, dandelion, purslane, or arugala)*

In a large salad bowl, mix together olive oil, vinegar, garlic, lemon juice, and oregano. Add wild greens and toss.

Note: By putting the dressing on the bottom of the bowl, you will find that your greens become lightly coated with flavor. Pouring the dressing on top tends to make them clumpy and soggy.

|  | Antioxidant Points | Omega-3 Points | Slow-Release Points | Fermentation Points |
|---|---|---|---|---|
| **Native Food Score** | 2 | 1 (from wild greens) | 0 | 1 (vinegar) |

For some time, one of my favorite weeknight meals has been a recipe for lentil stew with wild greens that is in Paula Wolfert's *Mediterranean Grains and Greens*. Then, when I was on Crete, I tasted a number of similar stews served both in tavernas and home kitchens. While Paula's recipe still serves as the foundation, it has changed a little to incorporate the variations I encountered on Crete.

This is a perfect dish for a weeknight meal, or you can dress it up with yogurt, lemon wedges, and a beautiful green salad and serve it to dinner guests.

*1 cup small, dark lentils (such as belugas)*

*8 cups chicken stock (or use water)*

*1 teaspoon salt*

*1 medium potato, peeled and sliced paper thin*

*1 cup sliced carrots*

*3 tablespoons olive oil*

*1 cup chopped onion*

*1 pound (1 packed quart) leafy greens (such as spinach, dandelion, arugula, kale, beet greens, or a mix)*

*¼ cup roughly chopped parsley leaves*

*1 tablespoon minced garlic*

*Plain yogurt and lemon wedges for garnish*

Wash the lentils. Place the lentils in saucepan and cover with stock (or water) and salt. Bring to a boil and skim off any foam on top. Add the potato and carrots, partially cover, and cook over medium heat for 20 minutes.

Meanwhile heat the olive oil in a large skillet and slowly brown the onions. While the onion is browning, wash, stem, and chop the greens. Add the parsley and garlic to the skillet and sauté for a minute or two, then stir in the greens and allow them to wilt, covered.

Scrape the contents of the skillet, including the oil, into the saucepan with lentils. Combine all the ingredients, then continue cooking covered for another 20 minutes, or until thick and soupy. Garnish with a drizzle of yogurt and serve with a lemon wedge.

VARIATION: For a thicker soup, use less broth/water and mash some of the lentils and potatoes.

|  | Antioxidant Points | Omega-3 Points | Slow-Release Points | Fermentation Points |
|---|---|---|---|---|
| **Native Food Score** | 3 | 1 | 1 | 1 (yogurt) |

. . . . . . . . . . . . . . . . . . . . . . . . . . . . . . . . . . . . . . . . . . . . . . . . . . . . . . . . . . .

# FISH STEWED WITH WILD GREENS
# AND LEEKS
*Serves 4 to 6*

. . . . . . . . . . . . . . . . . . . . . . . . . . . . . . . . . . . . . . . . . . . . . . . . . . . . . . . . . . .

This has become one of my standard fish recipes. I make it with whatever wild greens and white fish are fresh and in season. For the fish, I usually choose cod, sole, or sea bass. It is delicious with toasted bulgur and onions. (See recipe on page 272.)

*2 to 2½ pounds delicate white fish*

*Juice of 2 lemons or ½ cup dry white wine*

*1 teaspoon salt (more or less, to your taste)*

*Freshly ground pepper*

*¼ cup olive oil*

*2 cloves garlic, minced*

*2 leeks or 1 onion, chopped*

*6 cups wild greens, chopped (such as Swiss chard, spinach, kale, beet greens. If using a leaf with a woody stem like chard or kale, remove stems.)*

*2 tablespoons chopped fresh herbs (including lemon thyme, Greek oregano, basil)*

Marinate the fish in lemon juice, salt, and pepper for 1 hour before cooking.

Heat the olive oil in a skillet over medium heat. Cook garlic and leeks (or onions) until translucent, about 5 minutes.

Add the wild greens, toss, and cook until wilted.

Lay the fish over the greens, top with fresh herbs and lemon juice from marinade, cover with a lid, and turn the flame down to simmer. Cook 15 to 18 minutes or until the fish is tender and flakes with a fork.

To prepare leeks, first wash thoroughly to remove soil and sand. Trim the rootlets and a portion of the green tops, remove the outer layer, and then chop.

|  | Antioxidant Points | Omega-3 Points | Slow-Release Points | Fermentation Points |
|---|---|---|---|---|
| Native Food Score | 3 | 2 | 0 | 1 (wine) |

## CHICKEN AVGOLEMONO (IN LEMON SAUCE) WITH GRILLED PEPPERS

*Serves 4 to 6*

Variations of this one-pot lemony chicken dish are served up all over Crete. In this particular recipe, I'm including grilled green and red peppers. I love the golden yellow sauce with the flecks of green and red. However, you can substitute wilted wild greens for the peppers. Every Cretan cook knows that the trick to getting a perfect chicken is *siga siga*, or "slow slow." Patience and low heat are key. As tempted as you might be, don't try and speed this up by turning up the burner!

1 free-range fryer chicken cut into serving pieces and skinned at least 80 percent (It is okay to leave some of the skin on for flavor.)

¼ cup chickpea, barley, or whole-wheat flour

1 teaspoon each salt and pepper

¼ cup extra virgin olive oil

1 yellow or red onion, chopped very fine

3 cloves garlic, minced

1 tablespoon tomato paste (I buy it in tubes and keep it in the fridge as a stock item.)

2 lemons (add an extra lemon for a greater lemony flavor)

1 cup dry white wine (such as sauvignon blanc) or any red wine or brandy you have available

4 grilled bell peppers of any color, cut into bite-sized pieces, fresh or drained from a jar (See the sidebar for information on grilling peppers.)

2 tablespoons finely chopped fennel leaves (the frilly green part, also called "frond"), or ½ teaspoon fennel seeds

1 free-range egg

Rinse chicken and pat dry. Toss the chicken with the flour and salt and pepper until it is evenly coated.

In a shallow Dutch oven or large sauté pan, heat the olive oil over medium-high heat. Brown the chicken on both sides (approximately 5 minutes on each side), remove, and set aside.

Add the onion to the pan and cook until translucent. Add the garlic, tomato paste, and juice from one of the lemons, and combine. Add back the chicken and coat it with the onion mixture. Add the wine. Cover and simmer at a very low heat for 30 minutes. Add the peppers and the fennel leaves, cover again and simmer for another 20–25 minutes.

In a bowl whisk the egg. Continue whisking while you pour in the remaining lemon juice and a cup-full of the broth from the cooked chicken. Then pour the avgolemono (lemon) mixture into the chicken pot and gently stir it in. Turn off the heat and let the dish sit covered on the stove for 5–10 minutes in order for the sauce to thicken.

|  | Antioxidant Points | Omega-3 Points | Slow-Release Points | Fermentation Points |
|---|---|---|---|---|
| Native Food Score | 3 | 1 | 0 | 1 (wine) |

## Note on Grilling Peppers

Option 1:

On the stove top: Lay the peppers directly on your stove burner over a medium flame. (You need a gas stove). Roast each side for 4–6 minutes until the skin is black. Remove and let cool. Split open the pepper and rinse under cold water removing the black skin and the seeds.

Option 2:

In the broiler: Place peppers on a cooking sheet 8–10 inches from the broiler flame. Broil each side until blackened. Then clean as described above.

# BULGUR AND ONIONS

Serves 2 to 6

In Crete, bulgur, (cracked whole wheat) is a favorite grain. I find that its rich nutty flavor is a perfect complement to the sauces you are creating in your main dish, be it chicken, lamb, fish, or vegetable stew. From a health perspective, bulgur is an excellent choice since it is higher in nutrients and fiber than most other grains.

1 tablespoon extra virgin olive oil

1 small onion, halved and thinly sliced

2 garlic cloves, minced

1 cup cracked bulgur

1 cup low-sodium broth or water

In a sauté pan, heat the olive oil over medium heat. Add the onion and garlic and cook until onions are translucent, about 5 minutes. Add the bulgur and stir to coat. Add the broth and bring to a boil. Reduce the heat, cover, and simmer over low heat until the broth is absorbed and the bulgur is tender, about 10 minutes. While it is cooking, refrain from removing the cover or you will lose that valuable steam heat! For a finishing touch, remove the lid from the pan and toast the bulgur under the broiler for 2–3 minutes until the top grains are golden brown.

|  | Antioxidant Points | Omega-3 Points | Slow-Release Points | Fermentation Points |
|---|---|---|---|---|
| **Native Food Score** | 1 | 0 | 1 | 0 |

# SOFEGADA (SLOW-COOKED VEGETABLE STEW)

Serves 6 to 8

During my time on Crete, I had this stew on a number of occasions, each time with a different assortment of vegetables. Whenever I asked for the recipe, I was simply told to use a ¼ kilo of each type of vegetable. To make it easier for people who are not so comfortable in the metric system, let's just simplify and say a half pound of each vegetable. You will want to select at least four in-season vegetables from the following list for your stew. I usually serve this stew as a meze with a dark whole-grain bread or a barley rusk.

Note: When preparing the vegetables, the key is to not cut them too small or they will disintegrate with the slow cooking. (My standard rule of thumb is to make 1-inch slices or cubes.) You also want to avoid adding too much water or else the final stew will be soupy, and you will lose the rich flavor.

½ cup olive oil

2 onions, chopped

4 cloves garlic, minced

4 cups peeled and chopped tomatoes, or two 14-ounce cans of peeled, diced tomatoes (or a combination)

3 bell peppers, chopped

Vegetables to choose from:

    Small, waxy potatoes, sliced

Thin eggplants, sliced

Zucchini, sliced

Yellow squash, sliced

Small whole okra with the tough ends removed

Green beans, trimmed but not chopped

Whole, tender pea pods

Salt and pepper to taste

Water or chicken broth to cover

In a sauté pan, heat the olive oil and cook the onions until they are translucent. Add the garlic and stir for another 30–60 seconds. Transfer to the slow cooker or Dutch oven (a thick metal cooking pot with a tight-fitting lid), and add tomatoes, peppers, and remaining vegetables. Sprinkle with salt and pepper. Add enough liquid to barely cover the top of the vegetables.

For slow cooker: Cook for 6–8 hours on the low setting.

For Dutch oven: Bring to a boil. Simmer for 90 minutes over very low heat.

| | Antioxidant Points | Omega-3 Points | Slow-Release Points | Fermentation Points |
|---|---|---|---|---|
| **Native Food Score** | 3 | 0 | 1 | 0 |

## SLOW-COOKED LAMB FROM LASITHI    *Serves 4 to 6*

A delicious feast, but be sure not to skimp on the quality and cut of the lamb as it really matters in this recipe. I usually use lamb shanks, but any bone-in cut of lamb should work. This dish goes wonderfully with the sofegada or a grain and a green salad.

*¼ cup extra virgin olive oil*

*2 large red or yellow onions, sliced*

*2 cloves garlic, diced*

*¼ teaspoon fennel seeds*

*2–3 pounds stewing lamb, bone in, trimmed of fat (such as lamb shanks)*

*Salt and pepper to taste*

*1 large lemon, sliced*

*¾ cup water*

In a Dutch oven or a large pan with a tight lid, heat olive oil over medium heat, being careful not to smoke the oil. Cook onions, garlic, and fennel seeds for 5 minutes until onions are soft and lightly browned. Remove with a slotted spoon. Add lamb and brown on all sides over medium-high heat. Season with salt and pepper and add the lemon and water.

Simmer covered, over very low heat for 2 hours, or transfer to a slow cooker and cook for 8 hours on low. For a truly rich flavor, the lamb should cook in as dry a pot as possible without burning.

| | Antioxidant Points | Omega-3 Points | Slow-Release Points | Fermentation Points |
|---|---|---|---|---|
| **Native Food Score** | 1 | 1 (with free-range lamb) | 0 | 0 |

## LAYERED VEGETABLE PIE, OR *BIREIKI*     *Serves 4 to 6*

This is a classic Cretan dish. It takes a bit more time to prepare than most, so it is perfect for a weekend feast. This particular recipe is an adaptation of one that was given to me by Allison's in-laws, who are originally from the island. The key is to roast the vegetables first to give a nice carmelized flavor and color.

¼ cup olive oil

6 red creamer potatoes sliced very thin (about ¼–⅛ inch thick)

6 zucchinis or small eggplants, sliced

2 large heirloom tomatoes, sliced (or use 6 Roma tomatoes)

½ cup feta cheese

3 scallions (bottom part), chopped

2 whole garlic bulbs, peeled and chopped

3 tablespoons chopped Greek oregano or marjoram (optional)

½ cup shredded kasseri cheese, or any other melting cheese

Preheat oven broiler. Grease a cookie sheet with olive oil, spread out potatoes and zucchini slices, and brush with olive oil. Broil for 20 minutes; pay careful attention so that slices do not burn. Remove and let cool off. Turn oven down to 350°F.

Brush a shallow baking dish with about a tablespoon of the olive oil. Layer potato slices, zucchini, and tomatoes. Add about half the feta, scallions, garlic, and herbs, and repeat until you've used up all the vegetables and feta. Top the casserole with kasseri cheese. Bake at 350°F for 45 minutes.

|  | Antioxidant Points | Omega-3 Points | Slow-Release Points | Fermentation Points |
|---|---|---|---|---|
| **Native Food Score** | 3 | 1 | 1 | 1 (cheese) |

This succulent dessert is easy and attractive. You need a sharp knife to cut the figs without squashing them. When selecting your figs, try to choose ones that are not overly ripe. When figs are not in season, use these same toppings with other fruits, including melons, apples, oranges, and pears. For a change, I sometimes serve these figs on top of greens as a salad course.

16 fresh figs

½ cup Greek-style yogurt (see the following sidebar), goat cheese, or ricotta cheese

⅓ cup chopped walnuts (preferably toasted)

⅓ cup honey (optional)

Make 4 even lengthwise cuts around each fig. Place four figs on each plate, and spoon in about a teaspoon of yogurt or cheese into each cut. Sprinkle with walnuts and drizzle with honey just before serving.

|  | Antioxidant Points | Omega-3 Points | Slow-Release Points | Fermentation Points |
|---|---|---|---|---|
| Native Food Score | 1 | 1 | 0 | 1 (yogurt) |

## How to Make Greek-Style Yogurt

Greek yogurt is much thicker and creamier than American- or European-style yogurts. Here is a simple trick to turn your standard yogurt into Greek yogurt: Pour plain whole milk or reduced fat yogurt into a muslin bag or fine mesh strainer. Let the bag or strainer sit over a bowl to drain excess liquid for about 2 hours or until desired thickness is achieved. Enjoy!

## WHOLE-GRAIN RYE AND BUTTERMILK
## PANCAKES WITH BLUEBERRY SYRUP          *Serves 4 to 6*

This is one of my family's favorite breakfasts. I often mix up several batches at once and save them in the refrigerator so that we can even enjoy these nutritious pancakes on a busy weekday morning.

Rye is a slow-release, hardy grain that can withstand the intense cold. For this reason, it is one of the main staple grains traditionally eaten in Iceland. I love it because it gives these fluffy pancakes a deep, rich flavor. If you do not have rye flour, you can substitute with barley flour. If you do not have blueberries for the syrup, then drizzle your pancakes with honey or maple syrup.

*Pancakes:*

*1 cup ground whole rye*

*½ cup whole-wheat flour*

*1 tablespoon baking powder*

*Dash of salt*

*2 eggs, beaten (preferably omega-3 enriched)*

*1½ cups buttermilk (more if needed to thin batter), or see the following sidebar for buttermilk substitute*

*1 tablespoon butter, melted*

*Blueberry Syrup (optional):*

*2 cups fresh or frozen blueberries*

*2 tablespoons fresh lemon juice*

*½ cup honey*

Sift the dry ingredients in a large mixing bowl. In a small bowl, mix egg, buttermilk, and butter. Add to dry ingredients and mix well.

Heat a large pan to medium-high heat. Coat with one thin slice of butter. Scoop about ⅓ cup of batter onto pan and cook until bubbles appear on top and edges look cooked. Use less batter to make smaller 4-inch diameter pancakes. Flip and cook other side until golden brown. (Note: these pancakes are thick and may need some extra time to cook the middle. If the outside is getting too brown, turn down the heat to medium to cook longer.)

Put berries, lemon juice, and honey in saucepan and bring to a boil. Simmer for 10 minutes. Serve hot on top of rye pancakes.

| | Antioxidant Points | Omega-3 Points | Slow-Release Points | Fermentation Points |
|---|---|---|---|---|
| Native Food Score | 2 | 1 (with blueberries) | 1 | 1 (buttermilk) |

## Quick and Easy Buttermilk Substitute

If you don't have buttermilk in your refrigerator, use 1 cup milk plus 1 tablespoon white vinegar or lemon juice to replace 1 cup buttermilk. Let stand for 5 minutes.

## RYE BREAD (RUGBRAUO)                                    Serves 8

This dark, high-fiber bread is a cinch to assemble and only takes patience and time to bake. This preparation is a modern version of an old Icelandic tradition—steam-baking breads overnight by burying them in the earth next to a hot spring. Hearty with a touch of sweet, this rye bread makes a delicious breakfast bread when smeared with thick yogurt or soft cheese and layered with smoked salmon.

3 cups rye flour

1½ cups whole-wheat flour

2 teaspoons baking powder

½ teaspoon baking soda

1 teaspoon salt

½ cup maple syrup or honey

1 cup buttermilk (see sidebar for buttermilk substitute)

Preheat oven to 215°F. Mix the dry ingredients in a large bowl. Add the buttermilk and syrup.

Turn dough out onto a clean surface and knead until soft but not sticky. Add more buttermilk if needed. Divide the dough in half, and shape each half into a separate well-buttered loaf pan. *Be sure to seal each pan tightly with foil so that the steam cannot escape.* Place in the oven and bake for 10–12 hours. Alternatively, bake at 325°F for 3–4 hours. Slice with a serrated bread knife and serve warm with berry jam and butter or with cheese and nitrate-free smoked fish.

Note: The final product won't look like traditional loaves in the bread aisle at your market. This loaf does not rise much and is dense with a hard crust, but soft and moist inside. It also freezes well.

|  | Antioxidant Points | Omega-3 Points | Slow-Release Points | Fermentation Points |
|---|---|---|---|---|
| **Native Food Score** | 1 | 0 | 1 | 1 (buttermilk) |

## LOX SANDWICH ON RYE BREAD

*Serves 2*

Nearly every meal in Iceland includes some fish. This sandwich is great for breakfast or an afternoon snack, and you can make it using the homemade rye bread recipe. The yogurt in this recipe replaces the traditional *skyr*, a fermented nonfat dairy product native to Iceland.

*2–3 tablespoons plain yogurt (See sidebar for Greek yogurt on page 276.)*

*1 tablespoon chopped fresh dill*

*2 slices rye bread (from recipe above)*

*2 ounces smoked salmon*

*4–6 slices cucumber*

Mix yogurt and dill and spread on rye bread. Layer with smoked salmon and cucumber and serve open faced.

|  | Antioxidant Points | Omega-3 Points | Slow-Release Points | Fermentation Points |
|---|---|---|---|---|
| **Native Food Score** | 2 | 1 | 1 | 1 (yogurt) |

## SWEET AND SOUR CABBAGE *Serves 6*

Cabbage is one of the few vegetables that grows well in the short Icelandic summers, and it is found in most home gardens. This is a colorful, antioxidant-packed side dish or lunch that is tasty when served either hot or cold.

*1 small red cabbage, sliced into thin strips*

*1 apple, peeled, cored, and diced*

*1 cup blueberry juice, bilberry juice, or any deep colored, unsweetened juice*

*2–3 tablespoons red wine or apple cider vinegar*

*3 whole cloves or ½ teaspoon ground cloves*

*1 tablespoon butter*

*Sugar (or honey) and salt to taste*

Place cabbage and apple in a medium-sized saucepan. Stir in juice, vinegar, and cloves. Bring to a boil, then cover and simmer on low for 45 minutes, stirring every so often. Cook longer if you prefer it softer. Add butter, allow it to melt and then stir it into the cabbage mixture. Taste first, then add sugar or salt as desired. Serve hot with Fish and Potato Mash (see the following recipe) and rye bread, or serve cold on a sandwich.

|  | Antioxidant Points | Omega-3 Points | Slow-Release Points | Fermentation Points |
|---|---|---|---|---|
| **Native Food Score** | 3 | 0 | 0 | 1 (vinegar) |

## FISH AND POTATO MASH (STAPPA) *Serves 4*

This Icelandic standard combines two favorite foods: fish and potatoes. In my experience, this dish is a good way to introduce fish to "fish haters," since the fish taste is somewhat disguised by the potatoes. For those who like their food a little spicier, you can mash in some horseradish. For more color, try salmon instead of a white fish or replace the regular potatoes with sweet potatoes.

1 pound white fish (such as cod or sole)

1 pound waxy potatoes with skin on (such as fingerling), cut into 1 to 2-inch chunks

1 cup warm milk (more or less as needed)

1 tablespoon butter or cream (optional)

Salt and pepper to taste

Chives or parsley for garnish

Cook fish for 3–5 minutes in a pot of boiling water until the fish becomes flaky. Remove with a slotted spoon and drain. Flake with a fork, making sure to remove any bones. (The skin is full of omega-3s, but if you find it unappetizing, then you can remove it as well.)

Meanwhile cook potatoes in same boiling water until soft, about 20 minutes. Drain water from potatoes. Add flaked white fish, milk, and butter or cream and mash to desired texture. Add salt and pepper to taste. Serve hot scooped over Sweet and Sour Cabbage. (An ice-cream scoop can shape the mash into a pretty presentation on the plate.) Top with chives or parsley.

FISH CAKE VARIATION: Take a ½-cup scoop of the hash, flatten into the shape of a pancake and roll in bread crumbs and a dash of cayenne. Fry in a little bit of butter, until browned and crispy on the outside.

| | Antioxidant Points | Omega-3 Points | Slow-Release Points | Fermentation Points |
|---|---|---|---|---|
| Native Food Score | 0 | 1 | 1 | 0 |

## SPLIT PEA SOUP

Serves 4 to 6

This is an easy recipe and just what you want on a chilly day. In Iceland it is often served at Christmas and other holidays. While traditionally the meat is taken out last minute and served in a separate bowl, I prefer to shred it and add it to the soup. For vegetarians, you can make this recipe without the meat. I usually serve this soup with freshly baked Icelandic Rye Bread and a side of Sweet and Sour Cabbage.

1 cup dried yellow or green split peas

1 cup onion, chopped

1 tablespoon fresh thyme or 1 teaspoon dried

1½ pounds salted pork shoulder or smoked lamb shank, chopped into 5–6 pieces

2 cups total cubed root vegetables, including potatoes, carrots, turnips, and rutabagas

Salt and pepper to taste

Place the dried peas, onion, and thyme in a Dutch oven or soup pot and add 2 quarts water. Bring to a boil. Simmer partly covered for an hour. Occasionally skim off the foam from the top. (This first step can be done the night before.)

Add the meat to the pot and simmer for another 30 minutes. Add the root vegetables and simmer another 20–30 minutes until tender. Add additional water as needed to make your soup the desired consistency. Optional: remove 1–2 cups and puree in a blender to give the soup a creamier consistency.

Serve with one piece of meat in each bowl.

|  | Antioxidant Points | Omega-3 Points | Slow-Release Points | Fermentation Points |
|---|---|---|---|---|
| **Native Food Score** | 2 | 1 (with free-range meat) | 1 | 0 |

## FEAST

BAKED COD WITH THYME-LEMON BUTTER
SAUCE AND ROOT VEGETABLES                     *Serves 4*

This tasty fish dish makes use of the limited vegetables that grow in Iceland. Many Icelandic dishes call for boiling root vegetables, but if that doesn't strike your fancy, try lightly coating them with olive oil and roasting them.

6 carrots, peeled and chopped into 1-inch pieces

2 rutabagas, peeled and chopped into 1-inch pieces

¼ teaspoon salt

3 tablespoons butter

1 teaspoon grated lemon peel

Juice of 1 large lemon (about ¼ cup or more)

2 tablespoons chopped fresh thyme

1½ to 2 pounds cod or sole, cut into 4 fillets

Thyme sprigs for garnish

Put chopped carrots and rutabagas in a steamer basket with plenty of water below basket. Cover and steam vegetables for 15–20 minutes until tender. Sprinkle with salt and set aside in a bowl.

Preheat oven to 450°F.

To make the sauce melt the butter in saucepan over low heat. Stir in lemon zest, lemon juice, and thyme.

Place the fish fillets in a baking dish. Drizzle each fillet with about 1 tablespoon of the thyme-lemon-butter sauce. Spread the sauce around to coat above and below the fish fillets. Bake for 8–10 minutes until fish flakes easily with a fork.

On a platter, place the fish with warm vegetables; drizzle the vegetables with the sauce and garnish with thyme sprigs.

VARIATIONS:

- Add 2 tablespoons Dijon mustard to lemon-butter sauce.
- Try shaking dulse flakes (seaweed) over vegetables.
- Use red potatoes in place of carrots and rutabagas.

| | Antioxidant Points | Omega-3 Points | Slow-Release Points | Fermentation Points |
|---|---|---|---|---|
| Native Food Score | 2 | 2 | 1 | 0 |

BLUEBERRY (BILBERRY) SOUP                    Serves 4 to 6

This dessert may seem a little odd to those of us who are not native Icelanders, but it is an old-time recipe that makes use of the island's abundant summertime bilberries or bogberries. I think the whipped cream is essential, though in Iceland this soup is also topped with dried bread croutons.

2 cups blueberries, fresh or frozen

4 cups water

½ cup sugar or maple syrup

2 teaspoons lemon juice

2–3 tablespoons cornstarch mixed in 2 tablespoons cold water

Fresh cream, whipped (if desired), not sweetened

Place the berries and water in a saucepan and bring to a boil. Stir in the sugar or syrup and lemon juice and simmer for 10 minutes. Add the cornstarch-water mixture and continue to stir over heat until the soup thickens. Serve hot or cold in small bowls with a dollop of unsweetened whipped cream.

VARIATIONS: Try adding ¼ teaspoon ground cardamon to the whipped cream.

| | Antioxidant Points | Omega-3 Points | Slow-Release Points | Fermentation Points |
|---|---|---|---|---|
| Native Food Score | 2 | 0 | 0 | 0 |

## CAMEROON, WEST AFRICA

. . . . . . . . . . . . . . . . . . . . . . . . . . . . . . . . . . . . . . . . . . . . . . . . . . . . . . . . . . . . . . . . . . . . . . . . . . . . . . .

MILLET PORRIDGE                                                                      Serves 4

. . . . . . . . . . . . . . . . . . . . . . . . . . . . . . . . . . . . . . . . . . . . . . . . . . . . . . . . . . . . . . . . . . . . . . . . . . . . . . .

In West Africa, millet porridge is traditionally prepared by soaking the grain with sour milk for several days. As we discussed throughout this book, fermentation has numerous health benefits. Nonetheless, this process is time consuming, and the end result is not for everyone. Definitely an acquired taste! This recipe takes only minutes to make, but it is missing the healthy fermented bacteria. You can compensate for this by topping your porridge with cultured buttermilk, kefir, or yogurt.

| | |
|---|---|
| 1 cup millet | ¼ cup chopped nuts |
| ¼ cup dried fruit | 1 tablespoon honey |
| 2 cups water | ¼ cup buttermilk, kefir, or yogurt |

Soak millet and dried fruit in water overnight. The next day, heat the pot to a boil, then reduce heat to low and simmer for 20 minutes. Serve in bowls sprinkled with chopped nuts, honey, and top with buttermilk, kefir or yogurt.

| | Antioxidant Points | Omega-3 Points | Slow-release Points | Fermentation Points |
|---|---|---|---|---|
| Native Food Score | 1 | 0 | 1 | 1 (with yogurt or buttermilk) |

## OKRA WITH A KICK

*Serves 4*

Native to Africa, okra is also frequently found in traditional African American cuisine. This is a unique vegetable, with a gooey texture similar to the nopales cactus used in native Mexican cooking. Don't let that deter you; these pods with pop-in-your-mouth seeds are delicious and extremely nutritious—a half of a cup supplies significant amounts of calcium, folate, magnesium, and potassium.

| | |
|---|---|
| 1 pound okra, cut into 1-inch pieces | chopped (smaller chiles are hotter) |
| 1 tablespoon peanut oil, palm fruit oil or olive oil | ½ cup chopped peanuts |
| 1 red onion, chopped | 2 garlic cloves, minced |
| 1 hot chile pepper of your choice, seeds and inside ribs removed, finely | Salt to taste |

Bring a medium pot of water to a boil, and add the okra. Reduce heat and simmer the okra for 5 minutes. It should be bright green and still slightly crunchy. Drain in a colander and rinse with cool water.

Heat the oil in large sauté pan on medium heat. Add the red onion, chile, peanuts, and garlic. Sauté while stirring often until onion is translucent and soft, about 10 minutes. Add in the okra, stir until heated, add salt to taste and serve.

VARIATIONS: Add 1 cup fresh corn cut off the cob or 1 cup chopped tomatoes.

| | Antioxidant Points | Omega-3 Points | Slow-Release Points | Fermentation Points |
|---|---|---|---|---|
| Native Food Score | 3 | 0 | 1 | 0 |

## PILI PILI

*Makes about 2 cups*

This is an all-purpose Cameroonian salsa. Throughout West Africa, the word *pili pili* is used to refer to a variety of hot chiles.

2 cups diced tomatoes, either fresh or from a can

¼ cup finely chopped onion

1 clove garlic, minced

Juice of 1 medium lemon

½ teaspoon hot paprika

½ teaspoon grated ginger

1 teaspoon brown sugar or molasses

½ teaspoon cayenne pepper or 1 fresh habanero chile, minced

¼ teaspoon salt

Puree all the ingredients in a blender or mortar and pestle. Store in a covered jar in the refrigerator. Serve cold with any of the Cameroonian dishes. Delicious served over rice or millet.

| | Antioxidant Points | Omega-3 Points | Slow-Release Points | Fermentation Points |
|---|---|---|---|---|
| Native Food Score | 2 | 0 | 0 | 0 |

## MADAME'S CHICKEN PILI PILI *Serves* 4

Madame used to grill this chicken over an open fire. Alternatively, you can cook this chicken under the oven broiler. If you use the broiler method, I recommend that you use boneless cuts of chicken so that it will cook more evenly. Serve with plantains or millet and greens.

> 4 chicken breast halves or 8 chicken thighs, skin on
> 2 cups Pili Pili (page 286)

Marinate the chicken for at least 30 minutes (the more time, the better) in the *Pili Pili* sauce.

For the broiler: Preheat your oven's broiler at the highest setting. Place the chicken in a broiler pan on the highest shelf in the oven and cook for 7–10 minutes on each side.

For the barbeque: Grill the chicken over medium heat, turning often until the chicken is well done. Baste occasionally with more sauce.

VARIATION: You can also make this recipe with any mild-tasting white fish such as a trout or tilapia; however, cut the cooking time in half.

|  | Antioxidant Points | Omega-3 Points | Slow-Release Points | Fermentation Points |
|---|---|---|---|---|
| **Native Food Score** | 2 | 0 | 0 | 0 |

## COLLARD GREENS *Serves 2 to 4*

This recipe was given to me by Jay Foster. He, in turn, got it from his grand-mother Cora. It is very similar to many of the wild greens recipes that I collected in Cameroon. According to Jay, the trick to making scrumptious greens is to avoid making them soupy. Therefore, you need to cook them "as dry as possible."

Collard greens, like other cruciferous vegetables, release sulforaphane when chewed. In laboratory studies, sulforaphane enhances the body's anticancer defenses. Collards are also rich in a long list of nutrients: vitamins A, B, C, E, and K; potassium; zinc; iron; and calcium to name a few. One cup of cooked collards gives almost as much calcium as one cup of milk. The cumin and ginger in this recipe give just the right amount of spice to offset the collard's natural bitterness.

1 pound collard greens or any other available green

1 cup water or low-sodium chicken broth

1 tablespoon olive oil, peanut oil, or lard

2 cloves garlic, minced

½ teaspoon ground cumin

½ teaspoon grated fresh ginger

1 teaspoon lemon juice

Salt to taste

Wash the greens and remove the thick woody stems. Place them in a sauté pan with 1 cup of water or broth and bring to a boil. Cover the pan and turn down the heat. Simmer for 15–20 minutes until the greens are tender. Drain the greens, but reserve all the remaining liquid in a bowl. Heat the oil in the sauté pan and add the greens. Mix in the reserved water, the garlic, cumin, and ginger and cook uncovered at a low heat until the mixture is almost dry. Add lemon juice and salt to taste. Serve with chicken or plantains.

|  | Antioxidant Points | Omega-3 Points | Slow-Release Points | Fermentation Points |
|---|---|---|---|---|
| Native Food Score | 3 | 0 | 0 | 0 |

## BASIC TOASTED MILLET

*Serves 4*

Native to Africa, millet is a highly nutritious grain, and since it is gluten free, it is also an excellent choice for anyone who is gluten sensitive. The hull is indigestible to humans, and therefore, this grain is almost always sold hulled. The natural flavors in millet are enhanced by lightly toasting the grain before cooking. If you are interested in *safely* fermenting your millet, please see fermentation resources in Appendix G.

2 teaspoons lard, peanut oil, or palm fruit oil

1 cup millet

2 cups water

Salt to taste

Heat the lard or oil in a sauté pan over medium heat. Stir in the millet and toast until lightly brown. Add the water, cover, and bring to a boil. Cook over a low heat for 25 minutes until all the water is absorbed. Remove from the heat and let sit for 5 minutes; fluff with a fork.

|  | Antioxidant Points | Omega-3 Points | Slow-Release Points | Fermentation Points |
|---|---|---|---|---|
| Native Food Score | 1 | 0 | 1 | 0 |

## Gaining from Your Grains

Phytate, or phytic acid, is a substance that binds important nutrients (including iron, zinc, and calcium) in the husks of seeds, beans, nuts, and grains, and removing the phytate allows for better absorption of these nutrients.

Many traditional cooks soak their grains overnight in a bowl of water before preparing them, a practice that dramatically decreases the amount of phytate in the grains. Sprouting foods or combining them with a fermented food such as yogurt will also *decrease* the amount of phytate.

Here is an abbreviated (but still wonderful) version of *ndole* for those of us who don't have all day to prepare this delicious dish. It was given to me by Genevieve Hilpert, a nursing student from Cameroon who is now living in San Francisco. In Cameroon, they make this dish with bitterleaf. You can substitute in a wide variety of greens, including collard, spinach, dandelion, and kale. The shrimp or fish flakes are optional but add a lot to this medley of flavors. You can use the same bonito fish flakes that are used in the Okinawa recipes.

| | |
|---|---|
| 1 onion, chopped | 3 tablespoons peanut oil or palm fruit oil |
| 3 cloves garlic, minced | 1 pound cubed boneless beef or chicken or 1 block tofu |
| 2 teaspoons grated fresh ginger | |
| ⅛–¼ teaspoon hot cayenne pepper (optional) | ½ cup dried fish or shrimp flakes or 1½ cups whole shrimp, shelled, deveined, and chopped |
| 3 tomatoes, sliced, or one 14-ounce can chopped tomatoes | ¼ cup unsalted peanut butter or ¾ cup pumpkin or squash seeds, finely ground in the blender or a clean coffee grinder |
| 1 pound or 8 packed cups greens | |
| ¾ cup water | Salt and pepper to taste |

In a mortar and pestle or a blender, make a paste of the onion, garlic, ginger, cayenne pepper, and tomatoes.

Removed the thick, woody stems from the greens and wash thoroughly. Boil the greens in the water, and then drain and gently squeeze out the excess water. Finely chop the greens and set aside.

Heat the oil on medium heat in a sauté pan, making sure that the oil does not smoke. Add the meat and stir for 4–5 minutes until brown on all sides. Add the paste. Cover and continue stirring occasionally over low heat until sauce is a rich orange color, about 10 minutes.

Add the greens to the pan and combine with the meat mixture. Add the dried fish and enough of the peanut butter or ground seeds to thicken the stew. Cover and cook over low heat for 5 more minutes. Season with salt, cayenne pepper, and black pepper to taste. Serve over millet or brown rice.

| | Antioxidant Points | Omega-3 Points | Slow-Release Points | Fermentation Points |
|---|---|---|---|---|
| Native Food Score | 3 | 1 | 0 | 0 |

* * * * * * * * * * * * * * * * * * * * * * * * * * * * * * * * * * * * * * * * * *

## FRIED PLANTAINS AS A SNACK, SIDE DISH, OR SWEET

Serves 8

* * * * * * * * * * * * * * * * * * * * * * * * * * * * * * * * * * * * * * * * * *

*This is a wonderful fried indulgence for a special occasion. As plantains get increasingly ripe, their calorie load and glycemic index increases. Therefore, I recommend using the medium-ripe ones that are slightly soft but certainly not as soft as a banana. You can dust these with salt and have them as a snack or side dish, or you can drzzle them with honey and have them for dessert.*

4 medium ripe plantains, peeled
Peanut oil for frying

Cut the peeled plantains into 1-inch bite-sized pieces. Heat about ¼ inch of the peanut oil in a frying pan over medium-high heat. Add the plantains and cook on each side for 2 minutes until toasty brown. Be careful not to smoke the oil or burn the plantains.

Lay on a paper towel or bag to drain excess oil. Sprinkle the plaintains with salt or drizzle with honey and serve hot.

| | Antioxidant Points | Omega-3 Points | Slow-Release Points | Fermentation Points |
|---|---|---|---|---|
| Native Food Score | 0 | 0 | 1 | 0 |

## OKINAWA, JAPAN

Aside from the breakfast porridge and the Okinawan soba, all of these recipes were given to me by my friend Kenji's mother-in-law, Fujiko Miyahira. All of the ingredients for this section can be found in a traditional Asian market, and many of them are also available in regular grocery stores and natural food stores. Just be sure to read your labels since many of the commercially sold Asian foods have added chemicals such as artificial flavoring, colorants, and preservatives.

## BREAKFAST RICE PORRIDGE

*Serves 4 to 6*

This is a filling Okinawa breakfast and much tastier than instant hot cereals. I often make this at night and then reheat it the next morning. The Haiga rice is a white rice, but it still has its germ; therefore, it is much more nutritious than regular white rice. To find a store near you that sells Haiga, go to www.tamakimai.com/retails.html.

6 cups water

1½ cups short-grain brown rice or Haiga white rice

Condiments include chopped hard-boiled egg, scallions, cooked greens left over from the night before, brown sugar, toasted nuts or seeds, dried fruit, or fresh fruit.

The night before, bring the water to a boil in a 4-quart pot. Add the rice and return to a rolling boil, cover, and remove from the heat. Let stand on the counter overnight.

In the morning, place the covered pot over medium-high heat and cook, stirring occasionally until the porridge is creamy. You may need to add a little water or soy milk. When the rice is the desired consistency and warmth, serve in bowls with the condiments.

|  | Antioxidant Points | Omega-3 Points | Slow-Release Points | Fermentation Points |
|---|---|---|---|---|
| **Native Food Score** | 1 (if fruits and nuts are added) | 0 | 1 | 0 |

Though not typically what Americans think of for a morning meal, this is a popular Okinawan breakfast. It's also a nutritious afternoon snack. In general, white miso, which is sweeter, is eaten in the morning and darker miso is reserved for later in the day.

| | |
|---|---|
| 5 cups water | ¼ block firm or extra-firm tofu chopped into small bite-sized squares |
| ¼ cup dried seaweed, such as hijiki, wakame, or kelp, crumbled or torn | ½ cup miso paste |

In a small pot bring the water to boil. Turn off the heat. Immediately add the seaweed and the tofu. Remove 1 cup of the water and mix in the miso paste. Add this paste water mixture back to the main pot, cover, and let sit for 5 minutes.

| | Antioxidant Points | Omega-3 Points | Slow-Release Points | Fermentation Points |
|---|---|---|---|---|
| **Native Food Score** | 2 | 1 | 0 | 1 (miso) |

For most Okinawan cooks, their family's soba broth recipe is a carefully guarded secret. I tried to follow several recipes that I clipped out of local magazines but none tasted authentic. Finally, while teaching at Chubu Hospital, I met Yaeko Yara, and she was kind enough to share her recipe with me. To my delight, it tasted just right. Traditional Okinawan soba calls for pork, but I have since substituted chicken or shiitake mushrooms and have come up with some interesting variations.

The only challenge with this recipe is that the broth takes 1–2 hours to cook. (The longer it cooks, the richer the taste.) Sometimes I make the broth the night before, and then reheat it for serving the next day. An advantage to this method is that the pork fat congeals on the surface of the cooled broth and can be easily removed and discarded.

6 cups cold water

3 tablespoon plus 1 teaspoon soy sauce

1 quarter-sized piece fresh ginger

¼ cup dried fish flakes (Bonito flakes are the most common.)

1 pound meaty pork ribs, cut into 2-inch pieces (You can substitute with chopped bone-in chicken parts or 8 dried shiitake mushrooms.)

1⅓ tablespoons brown sugar

1 tablespoon plus 1 teaspoon sweet sake or mirin

1 tablespoon rice vinegar (or 2 tablespoons plus 1 teaspoon if not using sake)

One 8-ounce package soba noodles

(These can be buckwheat, but in Okinawa, soba noodles are traditionally flour-egg noodles.)

Condiments:

2 sheets nori, cut into confetti-sized strips, using sharp scissors

4 scallions, sliced in ½-inch lengths on a sharp diagonal

2 tablespoons sesame seeds, toasted or ground

1 small daikon radish, peeled and grated, or 2 tablespoons dried daikon

One 2-inch piece fresh ginger, peeled and grated

2 teaspoons wasabi powder mixed with enough warm water to form a soft paste

For the broth: Fill a pot with the cold water. Add 1 teaspoon of the soy sauce, the ginger, the fish flakes, and the ribs and bring to a boil. Turn down heat and simmer for 1–2 hours, occasionally skimming the fat off of the top.

Mix in the brown sugar, sake or mirin, rice vinegar, and the remainder of the soy sauce.

For the noodles: Meanwhile, bring a large pot of water to a boil and cook the soba to the desired doneness—about 5 minutes. Drain and rinse noodles in cold water, and gently swish the noodles with your hands.

To serve: Evenly distribute the soba in four bowls. Pour the broth over the noodles, and then garnish with the nori, scallions, sesame seeds, daikon, ginger, and spareribs. Allow each diner to add their own wasabi as desired.

|  | Antioxidant Points | Omega-3 Points | Slow-Release Points | Fermentation Points |
|---|---|---|---|---|
| Native Food Score | 3 | 2 | 1 | 1 (sake, vinegar) |

## NIGANA AEMONO SALAD (MIXED VEGETABLE SALAD)

This delicious salad is a great lunch, snack, or appetizer.

*1 large bunch of spinach or other leafy green*
*¼ cup crumbled medium or firm tofu*

*1 tablespoon miso paste*
*Sesame seeds, toasted, for garnish*

Boil a pot of water and dunk the spinach in the boiling water for about 5 seconds until the spinach becomes bright green. Pour into a strainer and rinse in cold water. Squeeze out the excess liquid and chop the spinach. In a bowl, combine the spinach, tofu, and miso paste. Mix and serve. Top with the toasted sesame seeds.

|  | Antioxidant Points | Omega-3 Points | Slow-Release Points | Fermentation Points |
|---|---|---|---|---|
| Native Food Score | 2 | 0 | 0 | 1 (miso) |

SEAWEED SALAD: Make the Nigana Aemono Salad using a 2- 3-ounce package of a fine dried seaweed such as *kurome* instead of spinach. Soak the seaweed for 20 minutes in a bowl of hot water and then drain. Squeeze out liquid and continue with the previous recipe.

|  | Antioxidant Points | Omega-3 Points | Slow-Release Points | Fermentation Points |
|---|---|---|---|---|
| Native Food Score | 2 | 1 | 0 | 1 |

Mung bean sprouts are not only delicious and crunchy, but they are also a good source of fiber, folate, and B vitamins.

5 cups mung bean sprouts

¼ cup rice vinegar

3 tablespoons soy sauce

1 tablespoon peanut oil or sesame oil

1 teaspoon sugar

1 teaspoon dried fish or bonito flakes (optional)

Boil a pot of water. Add the bean sprouts and boil for 1 minute. Drain and add vinegar, soy sauce, oil, sugar, and bonito flakes.

VARIATION: Make this using 2½ cups shredded daikon and 2½ cups shredded carrot.

|  | Antioxidant Points | Omega-3 Points | Slow-Release Points | Fermentation Points |
|---|---|---|---|---|
| **Native Food Score** | 2 | 1 | 0 | 1 (vinegar) |

## CHAMPURU (STIR-FRY)          *Serves 4 to 6*

Pork appears yet again in this classic Okinawan dish. However, you get just as tasty a result if you substitute the pork with an equal amount of boneless chopped chicken or 1 cup of fresh shiitake mushrooms. *Goya*, a bitter-tasting oddly-shaped gourd, is the typical vegetable used in champuru recipes. If you can locate *goya* in your local Asian market then give it a try. Otherwise, enjoy other in-season vegetables.

⅓ cup miso paste

2 tablespoons sake or water

2 tablespoons rice vinegar

1 teaspoon brown sugar or honey

2 tablespoons peanut oil

¼ pound pork shoulder, thinly sliced (optional)

1 block extra-firm tofu

1 tablespoon ground turmeric

6 cups chopped vegetables (such as broccoli, zucchini, carrots, or bok choy)

4 scallions or chives, chopped in 1-inch lengths

1 egg, beaten

Mix the miso paste, sake or water, rice vinegar, and brown sugar or honey and set aside. (Extra sauce can be saved in refrigerator for 2 weeks and used on the Nigana Aemono Salad, page 295.)

Heat 1 tablespoon of the oil in a wok or a frying pan over medium-high heat. Sauté bite-sized pieces of the pork until done. Separately crumble the tofu into large pieces and sprinkle with the turmeric. Add this into the frying pan and sauté with the pork. Remove from heat and place the mixture on a plate.

Heat another tablespoon of oil in the frying pan or wok and sauté the vegetables on medium-high heat. Important: stir-fry harder vegetables first (such as broccoli, cauliflower, and carrots) and softer vegetables last (such as zucchini and bok choy). Add the scallions or chives and cook for a minute or two longer. Put the cooked tofu and pork in the frying pan and gently mix with the vegetables and onions. Pour the beaten egg over the ingredients and mix until the egg is cooked. Pour the miso sauce over the ingredients and mix quickly. Remove from heat. Serve with Jushi Rice.

| | Antioxidant Points | Omega-3 Points | Slow-Release Points | Fermentation Points |
|---|---|---|---|---|
| **Native Food Score** | 4 | 0 | 0 | 1 (miso, sake) |

## JUSHI RICE

*Serves 6*

A favorite dish for Okinawan kids, this dish can easily be prepared in a pot or a rice cooker. While the traditional recipe calls for purple potatoes, you can throw in whatever root vegetables you have on hand. Jushi rice is a perfect accompaniment to the salads and the stir-fry dishes in this recipe section. For a special feast, serve it with the deep-fried fish.

2 cups short-grain brown rice or Haiga rice

1 purple potato, purple yam, or Okinawan sweet potato, cut in ¼-inch cubes

2 tablespoons low-sodium soy sauce (optional)

¼ cup dried shiitake mushrooms, stems removed, chopped, or 1 cup fresh shiitake mushrooms (optional)

2½ cups water

Mix rice, potato, soy sauce, and shiitake mushrooms in a pot, then add water and cover. Bring the covered pot to a boil and then turn down to a simmer for 50 minutes (check rice label for exact cooking time, which varies depending on the type of rice). Remove the lid and turn heat up for 10 seconds. If you making this in a rice cooker, use the 50 minute setting.

VARIATION: Use 1 cup black rice (forbidden rice) and 1 cup brown or white rice.

| | Antioxidant Points | Omega-3 Points | Slow-Release Points | Fermentation Points |
|---|---|---|---|---|
| Native Food Score | 1 | 0 | 1 | 1 |

• • • • • • • • • • • • • • • • • • • • • • • • • • • • • • • • • • • • • • • • • • • • • • • • • • •

## FEAST: OKINAWAN DEEP-FRIED FISH    *Serves 4*

• • • • • • • • • • • • • • • • • • • • • • • • • • • • • • • • • • • • • • • • • • • • • • • • • • • •

This is a serious treat that should only be made once a month because it is deep-fried. In Okinawa, they use *gurukun* (the official Okinawan fish), but I usually use trout. For more on deep-frying, see "*Deep-frying Responsibly*" sidebar in chapter 8. Serve with Nigana Aemono Salad or Seaweed Salad and Jushi Rice.

½ cup whole-wheat flour or brown rice flour

1 teaspoon salt

Two 8-inch long fish, split, about 1½ pounds

Peanut oil

Mix flour with salt. Dredge each fish entirely with flour on both sides. Heat about ¼ inch of peanut oil in a frying pan over medium-high heat. Fry fish for 3 minutes on each side (or until light brown), and then leave on a paper towel or paper bag for a minute or two to drain.

|  | Antioxidant Points | Omega-3 Points | Slow-Release Points | Fermentation Points |
|---|---|---|---|---|
| **Native Food Score** | 0 | 2 | 0 | 0 |

## IMO CUSTARD

*Serves 4 to 6*

A favorite Okinawan dessert, this may not seem very sweet to the Western palate. However, it offers a simple, nutritious meal-ender and is a good way to start to retrain taste buds that are accustomed to too much sugar.

4 medium-sized Okinawan sweet potatoes or purple yams, peeled, boiled, and mashed

3 tablespoons brown sugar

1 teaspoon grated ginger

Soy milk as needed

Using a hand masher or electric mixer, blend all ingredients until creamy. You can add soy milk until you get the desired consistency. Serve at room temperature in a bowl garnished with brown sugar.

|  | Antioxidant Points | Omega-3 Points | Slow-Release Points | Fermentation Points |
|---|---|---|---|---|
| **Native Food Score** | 2 | 0 | 1 | 0 |

# Appendix A: Cold Spot Shopping Lists

Here are shopping lists for each chapter of this book. They are arranged in such a way that they will match up with the shopping sections in most markets. The foods that have an asterisk (*) next to them are a little more unusual and may need to be purchased at an ethnic, natural, or specialty food store. Some of the seeds, rice, and grains can also be found online.

## COPPER CANYON, MEXICO

| Spices, Condiments, and Oils | Produce | Nuts, Grains, and Seeds | Meat, Fish, and Dairy |
|---|---|---|---|
| Cinnamon | Jicama | Masa harina* | Whole organic milk |
| Vanilla | Limes | Baby oats | Free-range pork or turkey bacon |
| Honey | Nopal cactus* | Spelt flakes* | Pacific cod or red snapper |

| Spices, Condiments, and Oils | Produce | Nuts, Grains, and Seeds | Meat, Fish, and Dairy |
| --- | --- | --- | --- |
| Ancho or habanero chile powder | Plum tomatoes | Dried red or black beans | Free-range eggs |
| Garlic | | Corn tortillas (without preservatives) | Queso fresco/blanco* |
| Epazote* | Serrano chiles, jalapeño chiles guajillo chiles,* and poblano chiles | Hominy or pozole* | Free-range or sustainably farmed whole chicken |
| Lard or olive oil | Oranges | | Trout |
| Achiote paste or annatto powder* | Tomatillos | | |
| Thyme | Banana leaves* | | |
| | Cilantro | | |
| | Red onions | | |
| | Cabbage | | |
| | Radishes | | |
| | Zucchini or other squash | | |
| | Green beans | | |
| | Corn | | |
| | Broad beans | | |

# CRETE, GREECE

| Spices, Condiments, and Oils | Produce | Nuts, Grains, and Seeds | Meat, Fish, and Dairy |
|---|---|---|---|
| Olive oil | Olives | Mini lentils | Feta cheese |
| Garlic | Tomatoes (heirloom) | Cracked bulgur* | Goat cheese |
| Fennel seeds | Oregano (Greek), basil, parsley, and marjoram | Whole-wheat flour or chickpea flour | Kasseri cheese* |
| Rose water or orange blossom water* | Fresh greens (kale, Swiss chard, purlsane, spinach, arugula, and so on) | Walnuts | Plain whole-milk yogurt |
| Sweet wine | Potatoes (russet and waxy) | Barley flour | Free-range eggs |
| Dry wine | Onions (red and yellow) | Yeast | Free-range or sustainably farmed chicken |
| Honey | Cilantro | | Lamb |
| Vinegar | Lemons | | Salmon, halibut, or cod |
| | Leeks | | |
| | Scallions | | |
| | Tomato paste | | |
| | Bell peppers | | |
| | Fennel | | |
| | Eggplant | | |
| | Zucchini or other squash | | |

| Spices, Condiments, and Oils | Produce | Nuts, Grains, and Seeds | Meat, Fish, and Dairy |
|---|---|---|---|
| | Okra | | |
| | Green beans | | |
| | Pea pods | | |
| | Raisins | | |
| | Figs | | |

# ICELAND

| Spices, Condiments, and Oils | Produce | Nuts, Grains, and Seeds | Meat, Fish, and Dairy |
|---|---|---|---|
| Honey | Sweet potatoes | Rye flour* | Free-range eggs |
| Maple syrup | Waxy potatoes | Whole-wheat flour | Whole organic milk |
| Red wine vinegar or black currant vinegar* | Lemons | Split peas (green or yellow) | Buttermilk |
| Cloves | Blueberries, fresh or frozen | | Butter, unsalted |
| Blueberry juice | Cabbage (red, purple, or green) | | White fish (cod or sole) |
| White wine vinegar | Apples | | Smoked salmon |
| Dill, fresh or dried | Cucumbers | | Halibut, wild salmon, or char |
| Bay leaves | Prunes | | Cream |
| Thyme, fresh or dried | Raisins | | Whole-milk Plain yogurt |

| Spices, Condiments, and Oils | Produce | Nuts, Grains, and Seeds | Meat, Fish, and Dairy |
| --- | --- | --- | --- |
| Sweet paprika | Lemons | | Salted pork shoulder or smoked lamb shank |
| | Parsley | | Mussels or clams |
| | Dulse seaweed flakes* | | |
| | Carrots | | |
| | Rutabagas | | |

## CAMEROON, WEST AFRICA

| Spices, Condiments, and Oils | Produce | Nuts, Grains, and Seeds | Meat, Fish, and Dairy |
| --- | --- | --- | --- |
| Ginger | Collard or mustard greens or kale | Peanuts | Free-range or sustainably farmed whole chicken or breast halves |
| Peanut oil or palm fruit oil | Tomatoes | Millet | Cubed chicken, beef, or shrimp |
| Ancho or habanero chile peppers | Plantains | Pumpkin seeds | |
| Hot paprkia | Sweet potatoes | | |
| Cayenne pepper | Okra | | |
| Cumin | Red onions | | |
| Garlic | Dried fruit | | |
| Dried shrimp or other fish | Lemons | | |

| Spices, Condiments, and Oils | Produce | Nuts, Grains, and Seeds | Meat, Fish, and Dairy |
|---|---|---|---|
| Young ginger* or regular ginger | Shiitake mushrooms (dried or fresh) | Buckwheat or whole-wheat soba noodles | Free-range or sustainably farmed pork spareribs, chopped |
| Daikon radish | | | Pork shoulder |
| Low-sodium soy sauce | Dried seaweed (kombu, dulse, hijiki, wakame, and arame*) | Tamaki Haiga rice,* black forbidden rice, or short-grain brown rice | Extra-firm tofu |
| Low-sodium tamari sauce | Asian mustard cabbage* | | Whole fish (trout or sardines) |
| Miso paste | Goya squash,* regular green or yellow squash | | |
| Turmeric | Scallions or chives | | |
| Sake* | Okinawa sweet potatoes* or purple potatoes or yams | | |
| Peanut oil | Spinach | | |
| Bonito flakes* or other dried fish or shrimp | Bean sprouts | | |
| Rice vinegar | Stir-fry vegetables (broccoli, carrots, bok choy, and so on) | | |
| Brown sugar or molasses | | | |

# Appendix B: How to Cook Indigenous Grains and Beans

| Native Grain (1 cup, dry) | Cooking Liquid (cups of water or broth) | Time to Cook (minutes) | Total Yield (cups) |
|---|---|---|---|
| Amaranth | 2½ | 20–25 | 2½ |
| Barley, hulled | 3 | 75 | 3½ |
| Barley, pearled (partly refined) | 3 | 50–60 | 3½ |
| Brown rice, long grain | 2 | 40–50 | 3 |
| Brown rice, short grain | 2 | 60–75 | 3 |
| Corn grits, coarse-ground | 4 | 20 | 2½–3 |

| Native Grain (1 cup, dry) | Cooking Liquid (cups of water or broth) | Time to Cook (minutes) | Total Yield (cups) |
| --- | --- | --- | --- |
| Couscous, whole grain | 1 | 5 | 2 |
| Kasha | 1¾ | 30 | 4 |
| Millet | 2 | 20–30 | 3–4 |
| Oats, steel-cut | 3–4 | 20 | 4 |
| Spelt | 3–4 | 60 | 3 |
| Teff | 3 | 10–20 | 3 |
| Triticale | 3 | 105 | 2½ |
| Wheat, bulgur | 1 | 10–15 | 2½ |
| Wild rice | 2½–3 | 45–55 | 4 |

The chart provides general guidelines that may differ depending on the brand of grain you purchase. Check recipes and package instructions for specifics. If a range is given for cooking liquid, the lower volume will give you a drier end product.

## Native Grain Preparation and Cooking Instructions

- *Soak, if desired*: Soaking grains overnight speeds the next day cooking time. It also makes the final product a bit softer and increases the digestibility of the grain's nutrients. It is convenient to do this with oats and rice for the breakfast meal.
- *Wash*: For most whole grains, especially quinoa, it's a good idea to wash them with water. This removes some of the saponins, a natural insecticide produced by the plant that can leave a bitter taste.
- *Toast, if desired*: For a nuttier flavor, after washing you can

toast grains for three to five minutes in a pan over medium heat until they are dry, but not burned. Cook in liquid as the recipe directs.

- *Stovetop steam cooking:* Bring water or broth to a boil, then stir in the grains, and bring back to a boil. Turn the heat to low and cover with a tight-fitting lid, simmering until the liquid is absorbed and grains are tender according to above cooking guidelines. Alternatively, you could put all the ingredients in the pot in cold water and then bring to a boil. However, boiling the water before adding the grains does a better job of keeping the grains separate.

- *Other grain-cooking methods:* You can also cook grains and beans in slow cooker, pressure cooker, countertop steamer, or rice cooker. Follow quantities and times in instruction booklets.

- *Hands off:*
  - Do not remove the lid while cooking. This allows the steam to escape and changes the cooking time.
  - Try not to stir grains while cooking; this has the effect of rupturing plant cell walls and making the grain mushy. After the grains become tender and the heat is turned off, let the grains sit covered for a few minutes before fluffing.
  - If grains are not fully cooked and water has been absorbed, add ¼ cup more water to finish steaming.

## Bean Cooking Chart

| Bean* (1 cup, dry) | Water (cups) | Cooking Time (minutes) |
|---|---|---|
| Adzuki | 4 | 45–55 |
| Anasazi | 3 | 45–55 |
| Black beans | 4 | 90 |

| Bean* (1 cup, dry) | Water (cups) | Cooking Time (minutes) |
| --- | --- | --- |
| Black-eyed peas | 3 | 60–75 |
| Cannellini (white) | 3 | 45 |
| Chickpeas (garbanzo) | 4 | 120–180 (2–3 hours) |
| Cranberry beans | 3 | 45 |
| Great Northern beans | 4 | 90 |
| Kidney beans | 3 | 90 |
| Lentils** | 2–3 | 30–45 |
| Lima beans, small | 4 | 50–75 |
| Lima beans, large | 4 | 60–90 |
| Navy beans | 3 | 60–90 |
| Pinto beans | 3 | 90 |
| Split peas** | 4 | 45–60 |

*1 cup of dried beans yields about 2–3 cups cooked
**Indicates bean does not need presoaking

## Bean Preparation and Cooking Instructions

- Wash and pick over beans for foreign objects, like pebbles and twigs.
- Most beans, with the exception of lentils and split peas, need to be soaked for at least eight hours or overnight to soften and make the beans more digestible. Soaking water should cover beans by three inches.
- Drain the soaking water and discard.
- Fill the pot with fresh water using amount and cooking times from the chart provided.
- Add salt and other seasonings *only* during the last ten minutes of cooking. Doing so beforehand prevents the bean skins from softening. Older dried beans and hard water also extend cooking time. So if possible, cook beans in filtered water.

- Add native herbs such as epazote sprigs, cumin, or bay leaves to the cooking beans to enhance flavor or reduce gassiness.
- To cook fresh beans (chickpeas, cranberry beans, black-eyed peas, and so on), remove from pods and either boil shelled beans or cover and steam in a steamer basket for five to ten minutes.

# Appendix C: More About Fats and Cooking Oils

It can be an overwhelming task to try to understand the nutritional facts about fats. Entire books have been dedicated to this topic and there is still much disagreement among experts in the field. The indigenous diets in this book teach us that a moderate amount of fat should be part of any healthy diet. There are a number of essential vitamins, such as vitamins A, D, E, and K, that are fat-soluble (meaning that they can only be absorbed with fatty foods), and certain fats play an important protective role in your health. The stories and recipes in this book will help you to figure out which fat-containing foods to eat and which to aviod. There are four main types of fats: trans or hydrogenated fats, saturated fats, monounsaturated fats, and polyunsaturated fats. These fats exist in different combinations and ratios in any given oil or fat-containing food.

- *Trans or hydrogenated fats* are denatured fats found in processed oils, margarines, and many baked goods. They are linked to cancer and heart disease and should be avoided. End of story.
- *Saturated fats*, from animal sources, are linked to heart disease and other chronic health problems, although none of these studies conclusively prove that it is the saturated fat (and not

some other factor) that is causing the health problem. Most of the calories in indigenous diets do not contain large amounts of saturated animal fats. I believe that the wisest approach is to emulate these eating patterns. Nonetheless, I do not think that these fats are as evil as many nutritionists claim. Recent research even suggests that whole-fat dairy products may offer some protective nutrients that are not found in nonfat or skim dairy. Saturated fats in certain indigenous vegetable oils (especially palm fruit and coconut oil) have also been shown to be beneficial when it comes to preventing chronic disease.

- Monounsturated (omega-9) fats such as fats from olives and avocados have many health benefits and make up a good proportion of fats found in indigenous diets. For more information about olive oil, see chapter 6.

- *Polyunstaurated fats* are essential fats. *Essential* means that our bodies cannot make them and need to get them from our diets. There are two types of essential fats: omega-3 fats (alpha linolenic acid, EPA or DHA) and omega-6 fats (linoleic acid). They get their "omega" name from the location of their double bonds in their molecular structures. A simple way to think of omega-3 fats is that they are the most *fluid* fats (they do not congeal in the fridge), and this is exactly how they behave in your body. They do not congeal in arteries, and they help maintain our cells and our nervous system. They also have an anti-inflammatory effect. Omega-6 fats are also important for our health, but they tend to be *stiffer* fats and cause more inflammation.

In all indigenous diets—where fats come from whole foods such as nuts, seeds, fruits, legumes, and wild game—omega-3 and omega-6 fats are found in about a 1:1 to 1:3 ratio. By contrast, the processed oils, margarines, and modern meats in our Western diets have shifted the balance away from omega-3 and monounsturated fats in favor of the more inflammatory trans, omega-6, and saturated fats. This has had drastic effects on our health, and many experts believe that this is a contributing factor to the rise of chronic diseases like heart disease, diabetes, rheumatoid arthritis, and cancer.

The refined vegetable oils that we use for cooking or that we eat in store-bought goods are one of the main causes of this omega-3 to omega-6 imbalance. These modern oils offer the advantage that they can tolerate high temperatures and have a long shelf life. They also have a neutral flavor and uniform color. However, they are a far cry from the fats used in indigenous cuisines around the world. In general, the cheaper brands of "refined" vegetable oils are extracted using chemical solvents (such as hexane, a toxic petroleum solvent). Once the oil is removed from the seed or pulp, it is then boiled to over 400°F to remove the toxic solvent. This is followed by bleaching and deodorizing; however, the final refined product often still has residues from the solvent. In addition, heating destroys the oil's natural antioxidants, including vitamin E and beta-carotene. This is why many refined oils contain synthetic antioxidants such as butylated hydroxytoluene (BHT). You can imagine that after learning about this process, I have made an effort to lessen my intake of these refined oils.

## What to Do with Your Used Oil

Cooking oil should never be dumped down the drain. A single gallon of oil can contaminate as much as a million gallons of water. With improper disposal of cooking oil, we can cause considerable harm to fish and other wildlife and destroy millions of gallons of potential drinking water. Cooking oil also congeals on our pipes, causing real plumbing disasters. The right way to dispose of your oil is to put it in a sealed nonrecyclable container and put it in with the regular garbage. Better yet, recycle it! Used cooking oil can be used to make soap and biodiesel fuel.

To find out where to recycle your oil, contact your local Household Hazardous Waste (HHW) disposal center.

With cold spot cooking, the majority of your fats should come from whole foods such as nuts, grains, seeds, legumes, fruits, and free-range meats. In my kitchen, when I use oil, my first choice is to use an indigenous, unrefined cooking fat with a high concentration of omega-9 fats or oleic acid and relatively little omega-6 fat. Extra virgin olive oil is an excellent choice. (See the Crete chapter for more detailed information on selecting your olive oil.) If extra virgin olive oil isn't a good match for the recipe, I usually use a teaspoon of butter or lard. For lower temperature Asian or West African recipes, my first choice is an unrefined peanut oil (ideally, organic).

On those very rare occasions when I decide to deep-fry or sear at high heats (370°F and above), I'll opt for refined peanut oil or palm fruit oil. But remember that it is possible to fry at a medium heat (350°F) and use olive oil or an unrefined oil so long as your oil is fresh and you do not put too much food in the pan at once. See chapter 8 for a more detailed discussion about deep-frying.

| No-Heat Oils/Fats | Medium-Heat Oils (up to 350°F) | Medium-High Heat Oils (up to 425°F) | High-Heat Oils (up to 510°F) |
|---|---|---|---|
| Nutritional supplement oils *(store in the fridge in a dark container!)* | • Sautéing, sauces, salad dressings, wok cooking <br>• Richer flavor | • Baking, sautéing | • Frying, sautéing <br><br>• Neutral flavor |
| Fish oil | Extra virgin olive oil | Butter | Refined avocado oil |
| Flaxseed oil | Unrefined sesame oil | Refined canola oil | Refined almond oil |
| Evening primrose oil | Unrefined peanut oil | Refined walnut oil | Refined super high-heat canola oil |

| No-Heat Oils/Fats | Medium-Heat Oils (up to 350°F) | Medium-High Heat Oils (up to 425°F) | High-Heat Oils (up to 510°F) |
|---|---|---|---|
| Borage seed oil | Butter | Refined peanut oil | Palm fruit oil |
| | Lard (up to 370°F) | | Refined sunflower oil |
| | Walnut oil | | Refined sesame oil |
| | | | Refined safflower oil (high oleic) |
| | | | Refined extra-light olive oil |

## Tips on Oil Storage and Use

- Store oils in a cool, dark place to protect from light, moisture, and oxygen. This will prevent oils from rapidly becoming rancid or developing a lower smoke point. Some unrefined oils can be refrigerated. Extra virgin olive oil (by nature not refined) does not require refrigeration, but must be used by its expiration date. If stored properly, you can keep a bottle of olive oil for up to two years—prior to the expiration date, of course.

- Avoid heating oil above its smoke point. When an oil starts to smoke or burn, it releases carcinogens in the air and forms dangerous free radicals that should not be ingested. Throw out any oil that smokes, turn on the stove fan, and start over.

- Do not reuse cooking oils used for frying. Each heating causes the fatty acids to break down and release harmful free radicals into the oil.

# Appendix D: Getting Your Omega-3 Fats

It is important to understand that foods contain omega-3 fats in a number of forms. The types of omega-3 fats that are most readily used by the body are those fats that are concentrated in cold-water fish, and to a lesser degree, in other free-range meats and eggs (namely DHA and EPA but also DPA). The omega-3 fats from wild greens, purslane, walnuts, and flax are ALA, or alpha-linolenic acid. Most people can convert at least a fraction of ALA to the more active DHA and EPA. However, if you stick to vegetarian sources of omega-3s, it usually takes a larger quantity of that given food to ensure that you have an adequate total amount of this vital nutrient.

My recommendation, which is based on looking at the indigenous cold spot diets and on current omega-3 research, is to have a minimum of 1.5 grams per day of omega-3 fats from vegetable or nonmarine sources, plus one to three servings per week of a low-mercury, environmentally friendly fish source. The indigenous food scores that accompany the recipes should serve as a guide to ensure that you are getting an adequate daily dose of omega-3s.

| Food Serving Size | Omega-3 (grams)* |
|---|---|
| *Low-Mercury Marine Sources* | |
| Cod liver oil, 1 tablespoon | 2.7 |
| Rainbow trout, 3 ounces | 0.7–1.0 |
| Salmon, wild, 3 ounces | 2.2 |
| Herring, 3 ounces | 1.9 |
| Sardines, 1 can, drained | 1.8 |
| Sardines, 1 can, in sardine oil | 5.5 |
| Salmon, canned, 3 ounces | 1.2 |
| Anchovies, 3 ounces | 1.2 |
| Caviar, 1 tablespoon | 1.1 |
| Striped bass (farmed), 3 ounces | 0.70 |
| *Nonmarine Meat and Dairy Sources* | |
| Free-range pork bacon, 3 medium slices | 0.50 |
| Free-range lamb, rib, roasted, 3 ounces | 0.50 |
| Free-range, omega-3–enriched egg | 0.175 |
| Organic milk, 1 cup | 0.175 |
| Alpine cheese, 3.5 ounces | 0.50 |
| *Major Nut, Seed, and Bean Sources* | |
| Soy nuts, ¼ cup | 0.75 |
| Walnut oil, 1 tablespoon | 1.4 |
| Walnuts, 1 ounce (14 halves) | 2.6 |
| Soy beans (green), boiled, 1 cup | 0.60 |
| Tofu, 4 ounces | 0.30 |
| Green beans, boiled, 1 cup | 0.60 |
| Flax seed oil, 1 tablespoon | 6.9 |
| *Green Sources* | |
| Green beans, boiled, 1 cup | 0.50 |
| Collard, kale, dandelion, mustard, or spinach, ½ cup, cooked | 0.10 |
| Seaweed, spirulina, dried 1 cup | 0.30 |
| Purslane, ½ cup, raw | 0.20 |

* Total of ALA, EPA, DPA (docosapentaenoic acid), and DHA

# Appendix E: Organic?
# A Priority List

Whenever possible, I suggest that you try to buy locally produced food since it usually tastes better, has fewer fungicides and more nutrients, uses less fuel, and supports your local economy. If you frequent a farmers' market, you'll be buying locally and have a wonderful opportunity to talk to farmers about how their foods are grown. Whenever it is not cost prohibitive, I also suggest buying organic fruits and vegetables, meats, and dairy. I realize that not everyone can buy all-local, all-organic, all the time. Therefore, it is important to have an organic priority list. The following are foods that tend to have the highest concentration of chemicals, including pesticides, antibiotics, and/or hormones when they are grown or raised conventionally.

*Environmental Working Group Shopper's Guide to Pesticides in Produce:*

| Worst Offenders (Try whenever possible to buy organic) | Cleanest Produce (Lowest in Pesticides) |
|---|---|
| Peaches | Onions |
| Apples | Avocados |
| Sweet bell peppers | Sweet corn, frozen |
| Celery | Pineapples |
| Nectarines | Mangos |
| Strawberries | Asparagus |
| Cherries | Sweet peas, frozen |
| Pears | Kiwi fruit |
| Grapes, imported | Bananas |
| Spinach | Cabbage |
| Lettuce | Broccoli |
| Potatoes | Papaya |

Source: Environmental Working Group (www.ewg.org). The produce ranking was developed by analysts at the not-for-profit Environmental Working Group (EWG) based on the results of nearly 43,000 tests for pesticides on produce collected by the U.S. Department of Agriculture and the U.S. Food and Drug Administration between 2000 and 2004. For a full list, go to www.foodnews.org.

### Dairy (Milk, Butter, Yogurt, Kefir)
- Buy organic as your first choice
- Ideally from a local farm

### Beef/Pork
- Buy locally raised, grass-fed, natural, sustainably farmed, or organic as your first choice. While all of these terms mean something slightly different, they will usually get you a piece of meat that is much healthier than one that is conventionally raised. (Note: most cattle ranchers do not feed all-grass diets due to winter weather conditions.)

### Chicken/Eggs
- Buy organic, free-range poultry and eggs from small local producers
- Ideally, fresh eggs from the farmers' market taste best. In addition to being a more humane way to raise chickens, this diet makes eggs and poultry a richer source of the beneficial omega-3 fatty acids.

# Appendix F: Useful Cooking Equipment for Preparing Indigenous Foods

- Tortilla press for making homemade corn tortillas
- Mortar and pestle for grinding spices and herbs
- Dutch oven with lid for all kinds of soups and stews
- Cast-iron pans
- Slow cooker
- Hand masher
- Sharp knife
- Cheesecloth
- Hand grater
- Wooden mixing spoons

# Appendix G: Recommended Indigenous Cookbooks

## Copper Canyon, Mexico

Cox, Beverly and Martin Jacobs. *Spirit of the Earth: Native Cooking from Latin America*. Stewart, Tabori and Chang, New York, 2001.

Divina, Marlene and Fernando Divina. *Foods of the Americas: Native Recipes and Traditions*. Ten Speed Press, Berkeley, CA, 2004.

Quintana, Patricia. *The Taste of Mexico*. Stewart, Tabori and Chang, New York, 1993.

## Crete, Greece

Kochilas, Diane. *The Glorious Foods of Greece*. HarperCollins, New York, 2001.

Salaman, Rena and Jan Cutler. *The Food and Cooking of Greece*. Anness Publishing, London, England, 2005.

Wolfert, Paula. *Mediterranean Grains and Greens*. HarperCollins, New York, 1998.

———. *Mediterranean Cooking*. Harper Collins, New York, 1994.

## Iceland

Rögnvaldardóttir, Nanna. *Icelandic Food and Cookery*. Hippocrene Books, New York, 2002.

## West Africa

Haffner, Dorinda. *A Taste of Africa: Traditional and Modern African Cooking*. Ten Speed Press, Berkeley, CA, 2002.

Harris, Jessica B. *The Africa Cookbook: Taste of a Continent*. Simon & Schuster, New York, 1998.

Ogunsanya, Dopke Lillian. *"My Cooking" West African Cookbook*. Dupsy Enterprises, Pflugerville, Texas, 1998.

Otoo, David and Tamminay Otoo. *Authentic African Cusine from Ghana*. Sankofa, Colorado Springs, CO, 1997.

Samuelsson, Marcus. *The Soul of a New Cuisine: A Discovery of the Foods and Flavors of Africa*. John Wiley & Sons, Hoboken, NJ, 2006.

## Okinawa

Willcox, Bradley D., Craig Willcox, and Makoto Suzuki. *The Okinawa Diet Plan*. Clarkson Potter, New York, 2004.

## Fermented Foods

Fallon, Sally. *Nourishing Traditions: The Cookbook That Challenges Politically Correct Nutrition and the Diet Dictocrats*. New Trends Publishing, Washington, DC, 2001.

Katz, Sandor Ellix. *Wild Fermentation: The Flavor, Nutrition, and Craft of Live-Cultured Foods*. Chelsea Green Publishing, White River Junction, VT, 2003.

Prentice, Jessica. *Full Moon Feast*. Chelsea Green Publishing, White River Junction, VT, 2006. www.wisefoodways.com

A good resource for culture starters to make your own fermented foods: www.fermentedtreasures.com

## Foraging for Wild Greens

Thayer, Samuel. *The Forager's Harvest: A Guide to Identifying, Harvesting, and Preparing Edible Wild Plants*. Forager's Harvest Press, Ogema, WI, 2006. Informative Web site with links to books, newsletters, and workshops on the art of picking wild plants: www.wildfoodadventures.com.

# Notes

## Chapter 1: Dining in the Cold Spots

### The Causes of Modern Chronic Diseases

Epping-Jordan, J. E., et al. "Preventing Chronic Diseases: Taking Stepwise Action." *The Lancet*, 2005, 366: 1,667–1,671.

Kanjilal, S., et al. "Socioeconomic Status and Trends in Disparities in Four Major Risk Factors for Cardiovascular Disease Among US Adults, 1971–2002." *Archives of Internal Medicine*, 2006, 166 (21): 2,348–2,355.

Mokdad, A. H., et al. "Actual Causes of Death in the United States, 2000." *Journal of the American Medical Association*, 2004, 291 (10): 1,238–1,245.

*National Institute of Health Report on Diabetes*, 2006. Web site: diabetes.niddk.nih.gov/dm/pubs/statistics/index.htm#7.

### Early Soy Exposure and Breast Cancer Prevention

Nagata, C., et al. "Associations of Mammographic Density with Dietary Factors in Japanese Women." *Cancer Epidemiology Biomarkers & Prevention*, 2005, 14 (12): 2,877–2,880.

Shu, X. O., et al. "Soyfood Intake During Adolescence and Subsequent Risk of

Breast Cancer Among Chinese Women." *Cancer Epidemiology Biomarkers & Prevention*, 2001, 10 (5): 483–488.

Wu, A. H., et al. "Adolescent and Adult Soy Intake and Risk of Breast Cancer in Asian-Americans." *Carcinogenesis*, 2002, 23 (9): 1,491–1,496.

## The ATBC Study Results

"The Effect of Vitamin E and Beta-carotene on the Incidence of Lung Cancer and Other Cancers in Male Smokers. The Alpha-Tocopherol, Beta-Carotene Cancer Prevention Study Group." *New England Journal of Medicine*, 1994, 330 (15): 1,029–1,035.

## Perception of Sweetness and Blood Glucose Levels

Melanson, K., et al. "Blood Glucose and Meal Patterns in Time-blinded Males After Aspartame, Carbohydrate and Fat Consumption, in Relation to Sweetness Perception." *British Journal of Nutrition*, 1999, 82: 437–446.

## More About the Weston Price Foundation

Fallon, S. *Nourishing Traditions*. Washington, DC: New Trends Publishing, 2001.

Price, W. *Nutrition and Physical Degeneration*. La Mesa, CA: The Price-Pottenger Nutrition Foundation, 1945.

Weston Price Foundation Web site: http://www.westonaprice.org.

## The Migration Effect: Globalization, Migration, and Diet

Gil, A., J. Vioque, and E. Torija. "Usual Diet in Bubis, a Rural Immigrant Population of African Origin in Madrid." *Journal of Human Nutrition and Dietetics*, 2005, 18 (1): 25–32.

Gimeno, S.G.A., et al. "Prevalence and Seven-year Incidence of Type II Diabetes Mellitus in a Japanese-Brazilian Population: An Alarming Public Health Problem." *Diabetologia*, 2002, 45 (12): 1,635–1,638.

Hawkes, C. "Uneven Dietary Development: Linking the Policies and Processes of Globalization with the Nutrition Transition, Obesity, and Diet-related Chronic Diseases." *Globalization and Health*, 2006, 2 (1): 4.

Marks, L. S., et al. "Prostate Cancer in Native Japanese and Japanese-American Men: Effects of Dietary Differences on Prostatic Tissue." *Urology*, 2004, 64 (4): 765–771.

Patel, J. V., et al. "Impact of Migration on Coronary Heart Disease Risk Factors: Comparison of Gujaratis in Britain and Their Contemporaries in Villages of Origin in India." *Atherosclerosis*, 2006, 185 (2): 297–306.

Schulz, L. O., et al. "Effects of Traditional and Western Environments on Prevalence of Type 2 Diabetes in Pima Indians in Mexico and the United States. *Diabetes Care*, 2006, 29 (8): 1,866–1,871.

# Chapter 2: Anatomy of an Indigenous Diet

## Interviews with Experts in the Chapter

Harriet V. Kuhnlein, PhD., Interview on January 24, 2007.
Paul Rozin, PhD., Interview on January 24, 2007.

## Maize and Nixtamalization

Coe, S. *America's First Cuisines*. Austin: University of Texas Press, 1994.

Katz, S. H., M. L. Hediger, and L. A. Valleroy. "Traditional Maize Processing Techniques in the New World." *Science*, 1974, 184 (4,138): 765–773.

Rosado, J. L., et al. "Calcium Absorption from Corn Tortilla Is Relatively High and Is Dependent upon Calcium Content and Liming in Mexican Women." *Journal of Nutrition*, 2005, 135 (11): 2,578–2,581.

## Imitation, Eating Patterns, and Food Preferences

Birch, L. L. "Effects of Peer Models' Food Choices and Eating Behaviors on Preschoolers' Food Preferences." *Child Development*, 1980, 51: 489–496.

Fallon, A. E., P. Rozin, and P. Pliner. "The Child's Conception of Food: The Development of Food Rejections with Special Reference to Disgust and Contamination Sensitivity." *Child Development*, 1984, 55 (2): 566–575.

Rozin, P. "Acquisition of Stable Food Preferences." *Nutrition Reviews*, 1990, 48 (2): 106–113; discussion 114–131.

Rozin, P. and L. Millman. "Family Environment, Not Heredity, Accounts for Family Resemblances in Food Preferences and Attitudes: A Twin Study. *Appetite*, 1987, 8 (2): 125–134.

## Eating Customs in Different Cultures

Milton, K., D. Knight, and I. Crowe. "Comparative Aspects of Diet in Amazonian Forest Dwellers." *Philosophical Transactions: Biological Sciences*, 2006, 334: 253–263.

Rappoport, L. *How We Eat: Appetite, Culture, and the Psychology of Food*. Toronto, Canada: ECW Press, 2003.

## The Health Benefits of Fasting

Benli Aksungar, F., et al. "Effects of Intermittent Fasting on Serum Lipid Levels, Coagulation Status, and Plasma Homocysteine Levels." *Annals of Nutrition and Metabolism*, 2005, 49 (2): 77–82.

Mattson, M. P. and R. Wan. "Beneficial Effects of Intermittent Fasting and Caloric Restriction on the Cardiovascular and Cerebrovascular Systems." *The Journal of Nutritional Biochemistry*, 2005, 16 (3): 129–137.

Sarri, K., et al. "Greek Orthodox Fasting Rituals: A Hidden Characteristic of the Mediterranean Diet of Crete." *British Journal of Nutrition*, 2004, 92: 277–284.

## The Eskimo Diet

Draper, H. "The Aboriginal Eskimo Diet in Modern Perspective." *American Anthropologist*, 1977, 79 (2): 309–316.

## What Your Meat Eats Matters

Eaton, S. B. and M. Konner. "Paleolithic Nutrition. A Consideration of Its Nature and Current Implications." *New England Journal of Medicine*, 1985, 312 (5): 283–289.

Mann, N. J., E. N. Ponnampalam, Y. Yep, A. J. Sinclair. "Feeding Regimes Affect Fatty Acid Composition in Australian Beef Cattle." *Asian Pacific Journal of Clinical Nutrition*, 2003, 12 (Suppl:S38).

Robbins, J. *Diet for a New America*. Walpole, NH: Stillpoint, 1987.

## About the Modern Imbalance of Omega-6 to Omega-3 Fats

Allport, S. *The Queen of Fats: Why Omega-3s Were Removed from the Western Diet and What We Can Do to Replace Them.* Berkeley, CA: University of California Press, 2006.

Simopoulos, A. "Evolutionary Aspects of Diet, the Omega-6/Omega-3 Ratio and Genetic Variation: Nutritional Implications for Chronic Diseases." *Biomedical Pharmacotherapy*, 2006, 60: 502–507.

## The Value of Fermented Foods

Aderiye, B. I. and S. A. Laleye. "Relevance of Fermented Food Products in Southwest Nigeria." *Plant Foods for Human Nutrition* (formerly *Qualitas Plantarum*), 2003, 58 (3): 1–16.

Blaut, M. and T. Clavel. "Metabolic Diversity of the Intestinal Microbiota: Implications for Health and Disease." *Journal of Nutrition*, 2007, 137 (3): 751S–755.

Kalliomaki, M., et al. "Probiotics During the First Seven Years of Life: A Cumulative Risk Reduction of Eczema in a Randomized, Placebo-Controlled Trial." *Journal of Allergy and Clinical Immunology*, 2007, 119 (4): 1,019–1,021.

McMaster, L. D., et al. "Use of Traditional African Fermented Beverages as Delivery Vehicles for *Bifidobacterium lactis* DSM 10140." *International Journal of Food Microbiology*, 2005, 102 (2): 231–237.

Ouwehand, A. C. "Antiallergic Effects of Probiotics." *Journal of Nutrition*, 2007, 137 (3): 794S–797.

## The Power of Spice

Jagetia, G. and B. Aggarwal. "'Spicing Up' of the Immune System by Curcumin." *Journal of Clinical Immunology*, 2007, 27 (1): 19–35.

Sherman, P. W. and J. Billing. "Darwinian Gastronomy: Why We Use Spices." *Bioscience*, 1999, 49 (6): 453–463.

Shishodia, S., G. Sethi, and B. B. Aggarwal. "Curcumin: Getting Back to the Roots." *Annals of the NY Academy of Science*, 2005, 1056 (1): 206–217.

## Dr. Rozin's Research on the Post-Ingestive Instinct

Rozin, P. "Specific Hunger for Thiamine: Recovery from Deficiency and Thiamine Preference." *Journal of Comparative Psychology*, 1965, 59: 98.

Rozin, P. and J. W. Kalat. "Specific Hungers and Poison Avoidance as Adaptive Specializations of Learning." *Psychology Reviews*, 1971, 78 (6): 459–486.

## Modern Food Additives

Griffiths, R. R. and E. M. Vernotica. "Is Caffeine a Flavoring Agent in Cola Soft Drinks?" *Archives of Family Medicine*, 2000, 9 (8): 727–734.

Guzelian, P. S., et al. "Evidence-Based Toxicology: A Comprehensive Framework for Causation." *Human & Experimental Toxicology*, 2005, 24: 161–201.

Soubra, L., et al. "Dietary Exposure of Children and Teenagers to Benzoates, Sulphites, Butylhydroxyanisol (BHA) and Butylhydroxytoluene (BHT) in Beirut (Lebanon)." *Regulatory Toxicology and Pharmacology*, 2007, 47 (1): 68–77.

Ward, M. H., et al. "Processed Meat Intake, CYP2A6 Activity, and Risk of Colorectal Adenoma." *Carcinogenesis*, 2007: bgm009.

# Chapter 3: A Diet Lost in Translation

## What Food Advertising Does to Our Food Choices

Nestle, M. *Food Politics*. Berkeley, CA: University of California Press, 2002.

Story, M. and S. French. "Food Advertising and Marketing Directed at Children and Adolescents in the United States." *International Journal of Behavioral Nutrition and Physical Activity*, 2004, 1 (1): 3.

Temple, J. L., et al. "Television Watching Increases Motivated Responding for Food and Energy Intake in Children." *American Journal of Clinical Nutrition*, 2007, 85 (2): 355–361.

## Health Effects of Losing the Family Dinner

Fitzpatrick, E., L. S. Edmunds, and B. A. Dennison. "Positive Effects of Family Dinner Are Undone by Television Viewing." *Journal of the American Dietetic Association*, 2007, 107 (4): 666–671.

Moreno, L. and G. Rodriguez. "Dietary Risk Factors for Development of Childhood Obesity." *Current Opinion in Clinical Nutrition and Metabolic Care*, 2007, 10 (3): 336–341.

Sen, B. "Frequency of Family Dinner and Adolescent Body Weight Status: Evidence from the National Longitudinal Survey of Youth, 1997." *Obesity*, 2006, 14 (12): 2,266–2,276.

## Sugary Drinks and Empty Calories

DellaValle, D. M., L. Roe, B. J. Rolls. "Does the Consumption of Caloric and Non-caloric Beverages with a Meal Affect Energy Intake?" *Appetite*, 2005, 44: 187–193.

Flood, J. E., L. Roe, B. J. Rolls. "The Effect of Increased Beverage Portion Size on Energy Intake at a Meal." *Journal of the American Dietetic Association*, 2006, 106: 1,984–1,991.

Mattes, R. D. "Beverages and Positive Energy Balance: The Menace is the Medium." *International Journal of Obesity* (Lond), 2006, 30: S60–65.

## Books on the Origins and History of Processed Grains

Diamond, J. *Guns, Germs, and Steel*. New York: W. W. Norton and Company, 1997.

Levenstein, H. *Paradox of Plenty: A Social History of Eating in Modern America*. Berkeley, CA: University of California Press, 2003.

## Manipulated Meat

Schlosser, E. *Fast Food Nation: The Dark Side of the All-American Meal*. Boston: Houghton Mifflin Co., 2001.

Wood, J. "Manipulating Meat Quality and Composition." *Proceedings of the Nutrition Society*, 1999, 58: 363–370.

# Chapter 4: Feeding Our Genes or Our Taste Buds?

## Experts Interviewed in This Chapter

Robert Nussbaum, MD, Telephone interview on February 13, 2007.
Jose Ordovas, PhD, Telephone interview on January 23, 2007.

Kauila Clark, MA, Telephone interview on December 19, 2006.
D. Craig Willcox, PhD, Personal communication, July 8, 2007.

## Returning to an Indigenous Diets

Jimenez-Cruz, A., et al. "A Flexible, Low-Glycemic Index Mexican-Style Diet in Overweight and Obese Subjects with Type 2 Diabetes Improves Metabolic Parameters During a Six-Week Treatment Period." *Diabetes Care*, 2003, 26 (7): 1,967–1,970.

Rowley, K., et al. "Improvements in circulating cholesterol, antioxidants, and homocysteine after dietary intervention in an Australian Aboriginal community." *American Journal of Clinical Nutrition*, 2001, 74: 442–448.

Shintani, T., et al. "Waianae Diet Program: Long-Term Follow-Up." *Hawaiian Medical Journal*, 1999, 58 (5): 117–122.

Tuomilehto, J. G., J. T. Salonen, A. Nissinen, K. Kuulasmaa, and P. Puska. "Decline in Cardiovascular Mortality in North Karelia and Other Parts of Finland." *British Medical Journal* 1986, 293: 1,068–1,071.

## Nutritional Adaptation and Food–Gene Interactions

Nabhan, G. P. *Why Some Like It Hot. Food, Genes, and Cultural Diversity.* Washington, DC: Island Press, 2004.

Ridley, M. *Genome: The Autobiography of a Species in Twenty-three Chapters.* New York: HarperCollins, 2000.

Stinson, S. "Nutritional Adaptation." *Annual Review of Anthropology*, 1992, 21: 143–170.

## Genetics Versus Environment: The Case of the Pima Indians

Schulz, L. O., et al. "Effects of Traditional and Western Environments on Prevalence of Type 2 Diabetes in Pima Indians in Mexico and the United States." *Diabetes Care*, 2006, 29 (8): 1,866–1,871.

## Family Medical History as a Risk Factor

Benhamiche-Bouvier, A. M., et al. "Family History and Risk of Colorectal Cancer: Implications for Screening Programmes." *Journal of Medical Screening*, 2000, 7: 136–140.

Khaw, K. and E. Barrett-Connor. "Family History of Heart Attack: A Modifiable Risk Factor?" *Circulation*, 1986, 74 (2): 239–244.

## The Waianae Diet Program

Shintani, T., et al. "The Waianae Diet Program: A Culturally Sensitive, Community-Based Obesity and Clinical Intervention Program for the Native Hawaiian Population." *Hawaiian Medical Journal*, 1994, 53 (5): 136–141, 147.

Shintani, T. T., et al. "The Hawaii Diet: Ad Libitum High-Carbohydrate, Low-Fat Multicultural Diet for the Reduction of Chronic Disease Risk Factors: Obesity, Hypertension, Hypercholesterolemia, and Hyperglycemia." *Hawaiian Medical Journal*, 2001. 60 (3): 69–73.

## Diet Swapping

Draper, H. "The Aboriginal Eskimo Diet in Modern Perspective." *American Anthropologist*, 1977, 79 (2): 309–316.

Kouris-Blazos, A., et al. "Are the Advantages of the Mediterranean Diet Transferable to Other Populations? A Cohort Study in Melbourne, Australia." *British Journal of Nutrition*, 1999, 82: 57–61.

Willcox, D. Craig, Okinawa, Japan. Interview, 2007.

# Chapter 5: Copper Canyon, Mexico: A Cold Spot for Diabetes

## Experts Interviewed in This Chapter

William Connor, MD, Telephone interview on January 11, 2006.
Maria Cruz, Interview on April 10, 2006.
Taurino, Interview on April 13, 2006.

## Diabetes in the United States

Engelgau, M. M., et al. "The Evolving Diabetes Burden in the United States." *Annals of Internal Medicine*, 2004, 140 (11): 945–950.

Gross, L. S., et al. "Increased Consumption of Refined Carbohydrates and the Epidemic of Type 2 Diabetes in the United States: An Ecologic Assessment." *American Journal of Clinical Nutrition*, 2004, 79 (5): 774–779.

Liao, Y., et al. "REACH 2010 Surveillance for Health Status in Minority Communities—United States, 2001–2002." *Morbidity and Mortality Weekly Report Surveillance Summary*, 2004, 53 (6): 1–36.

## Treating Prediabetes with Medication Versus Lifestyle

Diabetes Prevention Program Research. "Reduction in the Incidence of Type 2 Diabetes with Lifestyle Intervention or Metformin. *New England Journal of Medicine*, 2002, 346 (6): 393–403.

## Low-Glycemic Diets and Diabetes

Alfenas, R.C.G. and R. D. Mattes. "Influence of Glycemic Index/Load on Glycemic Response, Appetite, and Food Intake in Healthy Humans." *Diabetes Care*, 2005, 28 (9): 2,123–2,129.

Barclay, A. W., J. C. Brand-Miller, and T. M. S. Wolever. "Glycemic Index, Glycemic Load, and Glycemic Response Are Not the Same." *Diabetes Care*, 2005, 28 (7): 1,839–1,840.

Brand-Miller, J., et al. "Low-Glycemic Index Diets in the Management of Diabetes: A Meta-Analysis of Randomized Controlled Trials." *Diabetes Care*, 2003, 26 (8): 2,261–2,267.

Foster-Powell, K., S.H.A. Holt, and J. C. Brand-Miller. "International Table of Glycemic Index and Glycemic Load Values: 2002." *American Journal of Clinical Nutrition*, 2002, 76 (1): 5–56.

Hodge, A. M., et al. "Glycemic Index and Dietary Fiber and the Risk of Type 2 Diabetes." *Diabetes Care*, 2004, 27 (11): 2,701–2,706.

Wolever, T. M. S. and C. Mehling. "Long-Term Effect of Varying the Source or Amount of Dietary Carbohydrate on Postprandial Plasma Glucose, Insulin,

Triacylglycerol, and Free Fatty Acid Concentrations in Subjects with Impaired Glucose Tolerance." *American Journal of Clinical Nutrition*, 2003, 77 (3): 612–621.

## Slowly Digested Bush Foods

Thorburn, A. W., J. C. Brand, and A. S. Truswell. "Slowly Digested and Absorbed Carbohydrate in Traditional Bushfoods: A Protective Factor Against Diabetes?" *American Journal of Clinical Nutrition*, 1987, 45 (1): 98–106.

## Why Slow-Release Carbohydrates Are Antidiabetic

Manco, M., M. Calvani, and G. Mingrone. "Effects of Dietary Fatty Acids on Insulin Sensitivity and Secretion." *Diabetes, Obesity, and Metabolism*, 2004, 6 (6): 402–413.

Mann, N. J., "Paleolithic Nutrition: What Can We Learn from the Past?" *Asian Pacific Journal of Clinical Nutrition*, 2004, 13 (Suppl): S17.

Wolever, T.M.S. and C. Mehling. "Long-Term Effect of Varying the Source or Amount of Dietary Carbohydrate on Postprandial Plasma Glucose, Insulin, Triacylglycerol, and Free Fatty Acid Concentrations in Subjects with Impaired Glucose Tolerance." *American Journal of Clinical Nutrition*, 2003, 77 (3): 612–621.

## Diabetes Rates in Mexico and the U.S.–Mexico Border

Burke, J. P., et al. "Elevated Incidence of Type 2 Diabetes in San Antonio, Texas, Compared With That of Mexico City, Mexico." *Diabetes Care*, 2001, 24 (9): 1,573–1,578.

Rull, J. A., et al. "Epidemiology of Type 2 Diabetes in Mexico." *Archives of Medical Research*, 2005, 36 (3): 188–196.

## John Kennedy's Book on the Tarahumara

Kennedy, J. G., *Tarahumara of the Sierra Madre: Survivors on the Canyon's Edge*, 2nd ed. Pacific Grove, CA: Asilomar Press, 1996.

## Research on the Tarahumara Published by William Connor's Team

Cerqueira, M. T., M. M. Fry, and W. E. Connor. "The Food and Nutrient Intakes of the Tarahumara Indians of Mexico." *American Journal of Clinical Nutrition*, 1979, 32 (4): 905–915.

Connor, W., et al. "The Plasma Lipids, Lipoproteins, and Diet of the Tarahumara Indians of Mexico." *American Journal of Clinical Nutrition*, 1978, 31: 1,131–1,142.

McMurry, M., et al. "Changes in Lipid and Lipoprotein Levels and Body Weight in Tarahumara Indians After Consumption of an Affluent Diet." *New England Journal of Medicine*, 1991, 325: 1,704–1,708.

## The Pima and Type 2 Diabetes

Baier, L. J. and R. L. Hanson. "Genetic Studies of the Etiology of Type 2 Diabetes in Pima Indians: Hunting for Pieces to a Complicated Puzzle." *Diabetes*, 2004, 53 (5): 1,181–1,186.

Boyce, V. L. and B. A. Swinburn. "The Traditional Pima Indian diet. Composition and Adaptation for Use in a Dietary Intervention Study." *Diabetes Care*, 1993, 16 (1): 369–371.

Schulz, L. O., et al. "Effects of Traditional and Western Environments on Prevalence of Type 2 Diabetes in Pima Indians in Mexico and the United States." *Diabetes Care*, 2006, 29 (8): 1,866–1,871.

## History and Health Facts About Maize and Beans

Coe, S. *America's First Cuisines*. Austin: University of Texas Press, 1994.

Imungi, J. K. and J. N. Kabira. "Proximate Composition and Mineral Contents of Six Common Kenyan Varieties of Beans ('*Phaseolus vulgaris*')." *Ecology of Food and Nutrition*, 1989, 23: 13–16.

Landa-Habana, L., et al. "Effect of Cooking Procedures and Storage on Starch Bioavailability in Common Beans (*Phaseolus vulgaris*)." *Plant Foods for Human Nutrition* (formerly *Qualitas Plantarum*), 2004, 59 (4): 133–136.

Melo, E. A., et al. "Functional Properties of Yam Bean (*Pachyrhizus erosus*) Starch." *Bioresource Technology*, 2003, 89 (1): 103–106.

Roberts, J. *Cabbages and Kings: The Origins of Fruits and Vegetables*. London: Harper-Collins Publishers, 2001.

Sayago-Ayerdi, S. G., et al. "In-Vitro Starch Digestibility and Predicted Glycemic Index of Corn Tortilla, Black Beans, and Tortilla-Bean Mixture: Effect of Cold Storage." *Journal of Agriculture and Food Chemistry*, 2005, 53 (4): 1,281–1,285.

Wolever, T. M., D. J. Jenkins, L. U. Thompson, G. S. Wong, R. G. Josse. "Effect of Canning on the Blood Glucose Response to Beans in Patients with Type 2 Diabetes." *Human Nutrition-Clinical Nutrition*, 1987, 41: 135–140.

## The Hypoglycemic Effects of Nopales and Spices

Andrade-Cetto, A. and M. Heinrich. "Mexican Plants with Hypoglycaemic Effect Used in the Treatment of Diabetes." *Journal of Ethnopharmacology*, 2005, 99 (3): 325–348.

Laurenz, J. C., C. C. Collier, and J. O. Kuti. "Hypoglycaemic Effect of *Opuntia lindheimeri* Englem. in a Diabetic Pig Model." *Phytotherapy Research*, 2003, 17 (1): 26–29.

Trejo-Gonzalez, A., et al. "A Purified Extract from Prickly Pear Cactus (*Opuntia fuliginosa*) Controls Experimentally Induced Diabetes in Rats." *Journal of Ethnopharmacology*, 1996, 55 (1): 27–33.

## Low-Carbohydrate Diets and Diabetes

Boden, G., et al. "Effect of a Low-Carbohydrate Diet on Appetite, Blood Glucose Levels, and Insulin Resistance in Obese Patients with Type 2 Diabetes." *Annals of Internal Medicine*, 2005, 142 (6): 403–411.

McLaughlin, J., et al. "Adipose Tissue Triglyceride Fatty Acids and Atherosclerosis in Alaska Natives and Non-Natives." *Atherosclerosis*, 2005, 181 (2): 353–362.

## Vegan and Vegetarian Diets and Diabetes

Goff, L.M., et al. "Veganism and Its Relationship with Insulin Resistance and Intramyocellular Lipid." *European Journal of Clinical Nutrition*, 2005, 59 (2): 291–298.

Valachovičová, M. M., Krajčovičová-Kudláčková, P. Blažiček, and K. Babinská. "No Evidence of Insulin Resistance in Normal Weight Vegetarians." *European Journal of Nutrition*, 2006.

## High-Carbohydrate Native Diets and Diabetes

Gil, A., J. Vioque, and E. Torija. "Usual Diet in Bubis, a Rural Immigrant Population of African Origin in Madrid." *Journal of Human Nutrition and Dietetics*, 2005, 18 (1): 25–32.

Jimenez-Cruz, A., et al. "A Flexible, Low-Glycemic Index Mexican-Style Diet in Overweight and Obese Subjects with Type 2 Diabetes Improves Metabolic Parameters During a Six-Week Treatment Period." *Diabetes Care*, 2003. 26 (7): 1,967–1,970.

Rowley, K., et al. "Improvements in Circulating Cholesterol, Antioxidants, and Homocysteine After Dietary Intervention in an Australian Aboriginal Community." *American Journal of Clinical Nutrition*, 2001, 74: 442–448.

Shintani, T. T., S. Beckham, A. C. Brown, and H. K. O'Connor. "The Hawaii Diet: Ad Libitum High-Carbohydrate, Low-Fat Multicultural Diet for the Reduction of Chronic Disease Risk Factors: Obesity, Hypertension, Hypercholesterolemia, and Hyperglycemia." *Hawaiian Medical Journal*, 2001, 60: 69–73.

## Data on What America Eats

Centers for Disease Control, "Trends in Intake of Energy and Macronutrients—United States, 1971–2000." *Morbidity and Mortality Weekly Report*, 2004, 53: 80–82.

## The Slow-Release Effect of Fermented Foods

Leeman, M., E. Ostman, and I. Bjorck. "Vinegar Dressing and Cold Storage of Potatoes Lowers Postprandial Glycaemic and Insulinaemic Responses in Healthy Subjects." *European Journal of Clinical Nutrition*, 2005, 59 (11): 1,266–1,271.

Ostman, E. M., H. G. M. Liljeberg Elmstahl, and I. M. E. Bjorck. "Barley Bread Containing Lactic Acid Improves Glucose Tolerance at a Subsequent Meal in Healthy Men and Women." *Journal of Nutrition*, 2002, 132 (6): 1,173–1,175.

Ostman, E. M., H. G. M. Liljeberg Elmstahl, and I. M. E. Bjorck. "Inconsistency Between Glycemic and Insulinemic Responses to Regular and Fermented Milk Products." *American Journal of Clinical Nutrition*, 2001, 74 (1): 96–100.

## Animal Sources for Healthy Fats

Allport, S. *The Queen of Fats: Why Omega-3s Were Removed from the Western Diet and What We can Do to Replace Them*. Berkeley, CA: University of California Press, 2006.

Bee, G., G. Guex, and W. Herzog. "Free-range Rearing of Pigs During the Winter: Adaptations in Muscle Fiber Characteristics and Effects on Adipose Tissue Composition and Meat Quality Traits." *Journal of Animal Science*, 2004, 82 (4): 1,206–1,218.

Cordain, L., et al. "The Paradoxical Nature of Hunter-Gatherer Diets: Meat-Based, Yet Nonatherogenic." *European Journal of Clinical Nutrition*, 2002, 56 Suppl 1: S42–52.

Thorsdottir, I., J. Hill, and A. Ramel. "Omega-3 Fatty Acid Supply from Milk Associates with Lower Type 2 Diabetes in Men and Coronary Heart Disease in Women." *Preventive Medicine*, 2004, 39 (3): 630–634.

# Chapter 6: Crete, Greece: A Cold Spot for Heart Disease

## Experts Interviewed in This Chapter

Paul M. Ridker, MD, Telephone interview on March 28, 2006.

All other experts were interviewed on Crete from March 26 through April 1, 2007.

## Metabolic Syndrome

Isomaa, B. "A Major Health Hazard: The Metabolic Syndrome." *Life Sciences*, 2003, 73 (19): 2,395–2,411.

Jeppesen, J., et al. "Insulin Resistance, the Metabolic Syndrome, and Risk of Incident Cardiovascular Disease: A Population-Based Study." *Journal of the American College of Cardiology*, 2007, 49 (21): 2,112–2,119.

## Many Mediterranean Diets

Noah, A. and A. S. Truswell. "There Are Many Mediterranean Diets." *Asia Pacific Journal of Clinical Nutrition*, 2001, 10 (1): 2–9.

## Ancel Keys and the Seven Country Studies

Blackburn, H. "Ancel Keys." http://mbbnet.umn.edu/firsts/blackburn_h.html, Date accessed: March 20, 2007.

Keys, A. "Mediterranean Diet and Public Health: Personal Reflections." *American Journal of Clinical Nutrition*, 1995, 61: 1,321S–1,323S.

Keys, A. *Seven Countries: A Multivariate Analysis of Death and Coronary Heart Disease.* Cambridge MA: Harvard University Press, 1980.

Kromhout D., A. Keys, C. Aravanis, R. Buzina, et al. "Food Consumption Patterns in the 1960s in Seven Countries." *American Journal of Clinical Nutrition*, 1989, 49 (5): 889–894.

## Exporting the Mediterranean Diet

Hu, F. B. and W. C. Willett. "Optimal Diets for Prevention of Coronary Heart Disease." *Journal of the American Medical Association*, 2002, 288 (20): 2,569–2,578.

Kouris-Blazos, A., et al. "Are the Advantages of the Mediterranean Diet Transferable to Other Populations? A Cohort Study in Melbourne, Australia." *British Journal of Nutrition*, 1999, 82: 57–61.

Singh, R. B., et al. "Effect of an Indo-Mediterranean Diet on Progression of Coronary Artery Disease in High-Risk Patients (Indo-Mediterranean Diet Heart Study): A Randomised Single-Blind Trial." *The Lancet*, 2002, 360 (9,344): 1,455–1,461.

Waijers, P. M., et al. "Dietary Patterns and Survival in Older Dutch Women." *American Journal of Clinical Nutrition*, 2006, 83 (5): 1,170–1,176.

## Heart Disease and Inflammation

Ridker, P. M. "C-Reactive Protein and the Prediction of Cardiovascular Events Among Those at Intermediate Risk: Moving an Inflammatory Hypothesis Toward Consensus." *Journal of the American College of Cardiology*, 2007, 49 (21): 2,129–2,138.

Ridker, P. M. "C-Reactive Protein, Inflammation, and Cardiovascular Disease." *Current Issues in Cardiology*, 2005, 32 (3): 384–386.

Ridker, P. M., et al. "C-Reactive Protein Levels and Outcomes after Statin Therapy." *New England Journal of Medicine*, 2005, 352 (1): 20–28.

Ridker, P. M., et al. "Non-HDL Cholesterol, Apolipoproteins A-I and B100, Standard Lipid Measures, Lipid Ratios, and CRP as Risk Factors for Cardiovascular

Disease in Women." *Journal of the American Medical Association*, 2005, 294 (3): 326–333.

## Cholesterol–Lowering Drugs Versus. Heart Healthy Diets

Goldberg, A. C., et al. "Effect of Plant Stanol Tablets on Low-Density Lipoprotein Cholesterol Lowering in Patients on Statin Drugs." *The American Journal of Cardiology*, 2006, 97 (3): 376–379.

Jenkins, D. J. A., et al. "Assessment of the Longer-Term Effects of a Dietary Portfolio of Cholesterol-Lowering Foods in Hypercholesterolemia." *American Journal of Clinical Nutrition*, 2006, 83 (3): 582–591.

Jenkins, D. J. A., et al. "Direct Comparison of Dietary Portfolio Versus Statin on C-reactive Protein." *European Journal of Clinical Nutrition*, 2005, 59 (7): 851–860.

## The History of Crete

Detorakis, T. E. (John C. Davis, translator), *History of Crete*. Heraklion, Crete 1994.

## In Praise of Olive Oil

Andrikopoulos, N. K., et al. "Inhibitory Activity of Minor Polyphenolic and Non-polyphenolic Constituents of Olive Oil Against In Vitro Low-Density Lipoprotein Oxidation." *Journal of Medicinal Food*, 2002, 5 (1): 1–7.

Fito, M., et al. "Antioxidant Effect of Virgin Olive Oil in Patients with Stable Coronary Heart Disease: A Randomized, Crossover, Controlled, Clinical Trial." *Atherosclerosis*, 2005, 181 (1): 149–158.

Marrugat, J., et al. "Effects of Differing Phenolic Content in Dietary Olive Oils on Lipids and LDL Oxidation." *European Journal of Nutrition*, 2004, 43 (3): 140–147.

Visioli, F., et al. "Virgin Olive Oil Study (VOLOS): Vasoprotective Potential of Extra Virgin Olive Oil in Mildly Dyslipidemic Patients." *European Journal of Nutrition*, 2005, 44 (2): 121–127.

Weinbrenner, T., et al. "Olive Oils High in Phenolic Compounds Modulate Oxidative/Antioxidative Status in Men." *Journal of Nutrition*, 2004, 134 (9): 2,314–2,321.

## The Risks of Too Much Olive Oil

Menotti, A., et al., "Forty-Year Coronary Mortality Trends and Changes in Major Risk Factors in the First Ten Years of Follow-Up in the Seven Countries Study." *European Journal of Epidemiology*, 2007, 22(11): 747–754.

## The Other Plate Fellows: Foods That Complement the Olive Oil

Gorman, C. "More Than Just Olive Oil: The Benefits of Mediterranean Diets Are Real, but You Can't Drizzle Them on." Your Time/Health, 2003, 162 (1): 105.

Serra-Majem, L., et al. "Mediterranean Diet and Health: Is All the Secret in Olive Oil?" *Pathophysiology of Haemostasis and Thrombosis*, 2003, 33 (5–6): 461–465.

Serra-Majem, L., et al. "Olive Oil and the Mediterranean Diet: Beyond the Rhetoric." *European Journal of Clinical Nutrition*, 2003, 57 (9): S2(6).

Weisburger, J. H. "Lycopene and Tomato Products in Health Promotion." *Experimental Biology and Medicine*, 2002, 227 (10): 924–927.

## The Cardiovascular Benefits of Fasting

Sarri, K., et al. "Greek Orthodox Fasting Rituals: A Hidden Characteristic of the Mediterranean Diet of Crete." *British Journal of Nutrition*, 2004 92: 277–284.

Benli Aksungar, F., et al. "Effects of Intermittent Fasting on Serum Lipid Levels, Coagulation Status and Plasma Homocysteine Levels." *Annals of Nutrition and Metabolism*, 2005, 49 (2): 77–82.

## The Power of Greens

El, S. N. and S. Karakaya. "Radical Scavenging and Iron-Chelating Activities of Some Greens Used as Traditional Dishes in Mediterranean Diet. *International Journal of Food Sciences and Nutrition*, 2004, 55 (1): 67–74.

Ninfali, P., et al. "Antioxidant Capacity of Vegetables, Spices and Dressings Relevant to Nutrition." *British Journal of Nutrition*, 2005, 93 (2): 257–266.

Pitsavos, C., et al. "Adherence to the Mediterranean Diet Is Associated with Total

Antioxidant Capacity in Healthy Adults: The ATTICA Study." *American Journal of Clinical Nutrition*, 2005, 82 (3): 694–699.

Trichopoulo, A., E. Vasilopoulou, P. Hollman, et al. "Nutritional Composition and Flavonoid Content of Edible Wild Greens and Green Pies: A Potential Rich Source of Antioxidant Nutrients in the Mediterranean Diet." *Food Chemistry*, 2000, 70: 319–323.

Vasilopoulou, E., et al. "The Antioxidant Properties of Greek Foods and the Flavonoid Content of the Mediterranean Menu." *Current Medicinal Chemistry— Immunology, Endocrine & Metabolic Agents*, 2005, 5: 33–45.

## The FDA Definition of "Whole Grain"

FDA Web site: http://www.cfsan.fda.gov/~dms/flgragui.html. Accessed May 24, 2007.

## Slow-Release Carbohydrates and Heart Health

Mellen, P. B., T. F. Walsh, and D. M. Herrington. "Whole Grain Intake and Cardiovascular Disease: A Meta-Analysis." *Nutrition, Metabolism and Cardiovascular Diseases. American Journal of Clinical Nutrition* 2007 85 (6): 1444–5.

Wolever, T. M. S. and C. Mehling. "Long-Term Effect of Varying the Source or Amount of Dietary Carbohydrate on Postprandial Plasma Glucose, Insulin, Triacylglycerol, and Free Fatty Acid Concentrations in Subjects with Impaired Glucose Tolerance." *American Journal of Clinical Nutrition*, 2003, 77 (3): 612–621.

## Legumes for Cardiovascular Protection

Bazzano, L., et al. "Legume Consumption and Risk of Coronary Heart Disease in U.S. Men and Women: NHANES I Epidemiolgical Follow Up Study." *Archives of Internal Medicine*, 2001, 161: 2,573–2,578.

## Starches and Portion Size

Rozin, P., et al. "The Ecology of Eating: Smaller Portion Sizes in France Than in the United States Help Explain the French Paradox." *Psychological Science*, 2003, 14 (5): 450–454.

## Alcohol and the Heart

Bertelli, A. A., "Wine, research and cardiovascular disease: Instructions for use." *Atherosclerosis*. 2007 May 23.

Evert, J., et al. "Morbidity Profiles of Centenarians: Survivors, Delayers, and Escapers." *Journal of Gerontology Series A Biological Sciences and Medical Sciences*, 2003, 58 (3): M232–237.

Mukamal, K. J., S. E. Chiuve, and E. B. Rimm. "Alcohol Consumption and Risk for Coronary Heart Disease in Men with Healthy Lifestyles." *Archives of Internal Medicine*, 2006, 166 (19): 2,145–2,150.

Schroder, H., et al. "Myocardial Infarction and Alcohol Consumption: A Population-Based Case-Control Study." *Nutrition, Metabolism and Cardiovascular Diseases*, 2007 17 (8) 609–15.

## The Mediterranean Diet: A Study in Synergy

Hotz, C. and R. S. Gibson. "Traditional Food-Processing and Preparation Practices to Enhance the Bioavailability of Micronutrients in Plant-Based Diets." *Journal of Nutrition*, 2007, 137 (4): 1,097–1,100.

Trichopoulou, A., et al. "Adherence to a Mediterranean Diet and Survival in a Greek Population." *New England Journal of Medicine*, 2003, 348 (26): 2,599–2,608.

Trichopoulou, A., et al. "Mediterranean Diet in Relation to Body Mass Index and Waist-to-Hip Ratio: the Greek European Prospective Investigation into Cancer and Nutrition Study." *American Journal of Clinical Nutrition*, 2005, 82 (5): 935–940.

Trichopoulou, A., et al. "Traditional Foods: Why and How to Sustain Them." *Trends in Food Science and Technology*, 2006, 17: 498–504.

# Chapter 7: Iceland: A Cold Spot for Depression

## Experts Interviewed in This Chapter

Johann Axelsson, PhD, Interview in Reykjavik on July 20, 2006.

Joseph Hibbeln, MD, Telephone interview on June 2, 2006.

Richard Wurtman, MD, Telephone interview on June 2, 2006.

All other experts were interviewed in Iceland from July 20 through July 30, 2006.

## Depression Statistics and Information

WHO Report on Depression Worldwide. http://www.who.int/mental_health/management/depression/definition/en. Accessed February 23, 2007.

National Institute of Mental Health. http://www.nimh.nih.gov. Accessed February 23, 2007.

## The Links Between Food and Mood

Leigh Gibson, E. "Emotional Influences on Food Choice: Sensory, Physiological and Psychological pathways." *Physiology & Behavior.* 2006, 89 (1): 53–61.

Rozin, P., R. Bauer, and D. Catanese. "Food and Life, Pleasure and Worry, Among American College Students: Gender Differences and Regional Similarities." *Journal of Personality and Social Psychology,* 2003, 85 (1): 132–141.

## Iceland, A Depression Cold Spot

Copeland, J. E. A. "Depression Among Older People in Europe: The EURODEP Studies." *World Psychiatry,* 2004, February: 45–49.

Cott, J. and J. R. Hibbeln. "Lack of Seasonal Mood Change in Icelanders." *American Journal of Psychiatry,* 2001, 158 (2): 328.

Gudmundsson H., et al. "Inheritance of Human Longevity in Iceland. *European Journal of Human Genetics,* 2000, 8 (10): 743–749.

Lehtinen, V., et al. "The Estimated Incidence of Depressive Disorder and Its Determinants in the Finnish ODIN Sample." *Social Psychiatry and Psychiatric Epidemiology,* 2005, 40 (10): 778–784.

Magnusson, A. and D. Boivin. "Seasonal Affective Disorder: An Overview." *Chronobiology International,* 2003, 20 (2): 189–207.

Way, A., et al. "Obesity in Icelandic Youngsters and Canadian Youngsters of Icelandic Descent." *Arctic Medical Research,* 1988, 47 (Suppl 1): 462–465.

World Health Organization. "Suicide Rates per 100,000 by Country, Year, and Sex," *World Statistics 2007.* http://www.who.int/mental_heath/prevention/suicide_rates/en/.

## Depression, Testing the Gene Theory

Axelsson, J., R. Karadottir, and M. M. Karlsson. "Differences in the Prevalence of Seasonal Affective Disorder That Are Not Explained by Either Genetic or Latitude Differences." *International Journal of Circumpolar Health*, 2002, 61: 17–20.

Axelsson, J., et al. "Seasonal affective disorders: relevance of Icelandic and Icelandic-Canadian evidence to etiologic hypotheses." *Canadian Journal of Psychiatry*, 2002, (47) 2: 153–8.

Magnússon, J. and A. Axelsson. "The prevalence of seasonal affective disorder is low among descendants of Icelandic emigrants in Canada." *Archives of General Psychiatry*. 1993, 50 (12): 947–51.

Magnusson, A., J. Axelsson, M. M. Karlsson, and H. J. Oskarsson. "Lack of seasonal mood change in the Icelandic population: results of a cross-sectional study." *American Journal of Psychiatry*. 2000, 157 (2): 234–8.

## About Icelandic food

Rögnvaldardóttir, N. *Icelandic Food and Cookery*. New York: Hippocrene Books, 2002.

## Omega-3 Fats, Fish Consumption, and Depression

Hallahan, B. and M. R. Garland. "Essential Fatty Acids and Mental Health." *British Journal of Psychiatry*, 2005, 186 (4): 275–277.

Hibbeln, J. R. "Seafood Consumption, the DHA Content of Mother's Milk and Prevalence Rates of Postpartum Depression: A Cross-National, Ecological Analysis." *Journal of Affective Disorders*, 2002, 69 (1–3): 15–29.

Sontrop, J. and M. K. Campbell. "[Omega]-3 Polyunsaturated Fatty Acids and Depression: A Review of the Evidence and a Methodological Critique." *Preventive Medicine*. 2006, 42 (1): 4–13.

Tanskanen, A., et al. "Fish Consumption, Depression, and Suicidality in a General Population." *Archives of General Psychiatry*, 2001, 58 (5): 512–513.

Timonen, M., et al. "Fish Consumption and Depression: the Northern Finland 1966 Birth Cohort Study." *Journal of Affective Disorders*, 2004, 82 (3): 447–452.

## Fish Safety and Methylmercury

Cohen, J. T., et al. "A Quantitative Risk-Benefit Analysis of Changes in Population Fish Consumption." *American Journal of Preventive Medicine*, 2005, 29 (4): 325–334.

Egeland, G. M. and J. P. Middaugh. "Balancing Fish Consumption Benefits with Mercury Exposure." *Science*, 1997, 278 (5,345): 1,904–1,905.

Mozaffarian, D. and E. B. Rimm. "Fish Intake, Contaminants, and Human Health: Evaluating the Risks and the Benefits." *Journal of the American Medical Association*, 2006, 296 (15): 1,885–1,899.

## Omega–3 Content of Lamb and Organic Milk Products

Ellis, K. A., et al. "Comparing the Fatty Acid Composition of Organic and Conventional Milk." *Journal of Dairy Science*, 2006. 89 (6): 1,938–1,950.

Hauswirth, C. B., M. R. L. Scheeder, and J. H. Beer. "High {Omega}-3 Fatty Acid Content in Alpine Cheese: The Basis for an Alpine Paradox." *Circulation*, 2004, 109 (1): 103–107.

Mann, N. "Dietary Lean Red Meat and Human Evolution." *European Journal of Nutrition*, 2000, 39 (2): 71–79.

Nudda, A., et al. "Seasonal Variation in Conjugated Linoleic Acid and Vaccenic Acid in Milk Fat of Sheep and Its Transfer to Cheese and Ricotta." *Journal of Dairy Science*, 2005, 88 (4): 1,311–1,319.

Skuladottir, G., et al. "DHA in Maternal Plasma and Muscle Tissue of Newborn Lamb." In proceedings of Fourth International Congress on Essential Fatty Acids and Eicosanoids, 1997, Edinburgh, Scotland.

Thorsdottir, I., J. Hill, and A. Ramel. "Omega-3 Fatty Acid Supply from Milk Associates with Lower Type 2 Diabetes in Men and Coronary Heart Disease in Women." *Preventive Medicine*, 2004, 39 (3): 630–634.

## The Role of Antioxidants in Preventing Depression

Hintikka, J., et al. "Daily Tea Drinking Is Associated with a Low level of Depressive Symptoms in the Finnish General Population." *European Journal of Epidemiology*, 2005, 20 (4): 359–363.

Hintikka, J., et al. "High Vitamin $B_{12}$ Level and Good Treatment Outcome May Be Associated in Major Depressive Disorder." *Bio Med Central Psychiatry*, 2003, 3 (1): 17.

Somer, E. *Food and Mood: The Complete Guide to Eating Well and Feeling Your Best.* Henry Holt and Co., New York, 1999.

## Depression and Insulin Resistance

Liukkonen, T., et al. "The Association Between C-Reactive Protein Levels and Depression: Results from the Northern Finland 1966 Birth Cohort Study." *Biological Psychiatry*, 2006, 60 (8): 825–830.

Timonen, M., et al. "Depressive Symptoms and Insulin Resistance in Young Adult Males: Results from the Northern Finland 1966 Birth Cohort." *Molecular Psychiatry*, 2006, 11: 929–933.

Timonen, M., et al. "Insulin Resistance and Depression: Cross-Sectional Study." *British Medical Journal*, 2005, 330 (7,481): 17–18.

## Carbohydrates and Mood

Benton, D. "Carbohydrate Ingestion, Blood Glucose and Mood." *Neuroscience & Biobehavioral Reviews*, 2002, 26 (3): 293–308.

de Castro, J. M. "Macronutrient Relationships with Meal Patterns and Mood in the Spontaneous Feeding Behavior of Humans." *Physiology & Behavior*, 1987, 39 (5): 561–569.

Schweiger, U., R. Laessle, and K. Pirke. "Macronutrient Intake and Mood During Weight Reducing Diets. *Annals of the NY Academy of Sciences*, 1987, 449: 335–337.

Wurtman, J. J. "Depression and Weight Gain: the Serotonin Connection." *Journal of Affective Disorders*, 1993, 29 (2–3): 183–192.

Wurtman, R. J., et al. "Effects of Normal Meals Rich in Carbohydrates or Proteins on Plasma Tryptophan and Tyrosine Ratios." *American Journal of Clinical Nutrition*, 2003, 77 (1): 128–132.

## All About Spuds

Foster-Powell, K., S. H. A. Holt, and J. C. Brand-Miller. "International Table of Glycemic Index and Glycemic Load Values: 2002." *American Journal of Clinical Nutrition*, 2002, 76 (1): 5–56.

Leeman, M., E. Ostman, and I. Bjorck. "Vinegar Dressing and Cold Storage of Potatoes Lowers Postprandial Glycaemic and Insulinaemic Responses in Healthy Subjects." *European Journal of Clinical Nutrition*, 2005, 59 (11): 1,266–1,271.

van Dam, R. M., et al. "Patterns of Food Consumption and Risk Factors for Cardiovascular Disease in the General Dutch Population." *American Journal of Clinical Nutrition*, 2003. 77 (5): 1,156–1,163.

Waijers, P. M., et al. "Dietary Patterns and Survival in Older Dutch Women." *American Journal of Clinical Nutrition*, 2006, 83 (5): 1,170–1,176.

## The Benefits of a Slow-Release Breakfast

Benton, D., O. Slater, and R. T. Donohoe. "The Influence of Breakfast and a Snack on Psychological Functioning." *Physiology & Behavior*, 2001, 74 (4–5): 559–571.

Farshchi, H. R., M. A. Taylor, and I. A. Macdonald. "Deleterious Effects of Omitting Breakfast on Insulin Sensitivity and Fasting Lipid Profiles in Healthy Lean Women." *American Journal of Clinical Nutrition*, 2005, 81 (2): 388–396.

Mahoney, C. R., et al. "Effect of Breakfast Composition on Cognitive Processes in Elementary School Children." *Physiology & Behavior*, 2005, 85 (5): 635–645.

Nabb, S. and D. Benton. "The Influence on Cognition of the Interaction Between the Macro-Nutrient Content of Breakfast and Glucose Tolerance." *Physiology & Behavior*, 2006, 87 (1): 16–23.

Pasman W. J., et al. "Effect of Two Breakfasts, Different in Carbohydrate Composition, on Hunger and Satiety and Mood in Healthy Men." *Obesity and Related Metabolic Disorders*, 2003, 27 (6): 663–668.

Warren, J. M., C.J.K. Henry, and V. Simonite. "Low Glycemic Index Breakfasts and Reduced Food Intake in Preadolescent Children." *Pediatrics*, 2003, 112 (5): e414–.

## Chocolate and Mood

Rozin, P., E. Levine, and C. Stoess. "Chocolate Craving and Liking." *Appetite*, 1991, 17 (3): 199–212.

## Alcohol and Appetite

Hetherington, M. M., et al. "Stimulation of Appetite by Alcohol." *Physiology & Behavior*, 2001, 74 (3): 283–289.

# Chapter 8: Cameroon, West Africa: A Cold Spot for Bowel Trouble

## Experts Interviewed in This Chapter

James C. McCann, PhD, Telephone interview on April 28, 2006.
Stephen O'Keefe, MD, Telephone interview on May 9, 2006.
Jessica B. Harris, PhD, Telephone interview on July 11, 2006.
Jay Foster, Interview on November 21, 2006.

## Epidemiology of Colon Cancer in Africa and Around the World

Anderson, W., A. Umar, and O. Brawley. "Colorectal Carcinoma in Black and White Race." *Cancer and Metastasis Review*, 2003, 22: 67–82.

Burkitt, D. "Epidemiology of Cancer of the Colon and Rectum." *Cancer*, 1971, 28: 3–13.

Jemal, A., et al. "Cancer Statistics, 2006." *CA: A Cancer Journal for Clinicians*, 2006, 56 (2): 106–130.

Parkin, D. M., P. Pisani, and J. Ferlay. "Estimates of the Worldwide Incidence of Twenty-Five Major Cancers in 1990." *International Journal of Cancer*, 1999, 80 (6): 827–841.

Parkin, D. M., et al. "Global Cancer Statistics, 2002." *CA: A Cancer Journal for Clinicians*, 2005, 55 (2): 74–108.

Segal, I., C. A. Edwards, and A. R. P. Walker. "Continuing Low Colon Cancer Incidence in African Populations." *The American Journal of Gastroenterology*, 2000, 95 (4): 859–860.

## The "Bran Man," Denis Burkitt, MD

Coakley, D. "Denis Burkitt and His Contribution to Haematology/Oncology." *British Journal of Haematology*, 2006, 135 (1): 17–25.

Lee, T. "Seeing the Wood for the Trees—the Early Papers of Denis Burkitt." *Journal of Irish College of Physicians and Surgeons*, 1996, 25 (2): 126–130.

Nelson, E. "Out of Africa . . . Major Medical Discoveries—Dr. Denis Burkitt." *Saturday Evening Post*, 1995.

Trowell H. C. and D. P. Burkitt, eds. *Western Diseases: Their Emergence and Prevention.* Cambridge: Harvard University Press, 1981.

## Loss of Indigenous African Foods

Lappe, F. M. and A. Lappe. *Hope's Edge.* New York: Penguin, 2003.

Wylie, D. *Starving on a Full Stomach: Hunger and the Triumph of Cultural Racism in Modern South Africa.* Charlottesville: University of Virginia, 2001.

## Fiber, the First *F* in Colon Disease Prevention

Alonso-Coello, P., et al. "Fiber for the Treatment of Hemorrhoids Complications: A Systematic Review and Meta-Analysis." *The American Journal of Gastroenterology,* 2006, 101 (1): 181–188.

Baron, J. A. "Dietary Fiber and Colorectal Cancer: An Ongoing Saga." *Journal of the American Medical Association,* 2005, 294 (22): 2,904–2,906.

Bingham, S. A., et al. "Dietary Fiber in Food and Protection Against Colorectal Cancer in the European Prospective Investigation into Cancer and Nutrition (EPIC): An Observational Study." *The Lancet,* 2003, 361 (9,368): 1,496–1,501.

Korzenik, J. R. "Case Closed?: Diverticulitis: Epidemiology and Fiber." *Journal of Clinical Gastroenterology,* 2006, (Supplement 3): S112–S116.

Potter, J. D. "Fiber and Colorectal Cancer–Where to Now?" *New England Journal of Medicine,* 1999, 340 (3): 223–224.

## Less Flesh, the Second *F* in Colon Disease Prevention

Butler, L. M., et al. "Heterocyclic Amines, Meat Intake, and Association with Colon Cancer in a Population-Based Study." *American Journal of Epidemiology,* 2003, 157 (5): 434–445.

Edenharder, R., S.J., H. Glatt, E. Muckel, K. L. Platt. "Protection by Beverages, Fruits, Vegetables, Herbs, and Flavonoids Against Genotoxicity of 2-Acetyl-aminofluorene and 2-Amino-1-Methyl-6-Phenylimidazo[4,5-b]Pyridine (PhIP) in Metabolically Competent V79 Cells." *Mutation Research,* 2002, 521: 57–72.

Edenharder, R., A. Worf-Wandelburg, M. Decker, and K.L. Platt. "Antimutagenic Effects and Possible Mechanisms of Action of Vitamins and Related Compounds

Against Genotoxic Heterocyclic Amines from Cooked Food." *Mutation Research*, 1999, 444: 235–248.

English, D. R., et al. "Red Meat, Chicken, and Fish Consumption and Risk of Colorectal Cancer." *Cancer Epidemiological Biomarkers & Prevention*, 2004, 13 (9): 1,509–1,514.

Mirvish, S. S., et al. "Total N-Nitroso Compounds and Their Precursors in Hot Dogs and in the Gastrointestinal Tract and Feces of Rats and Mice: Possible Etiologic Agents for Colon Cancer." *Journal of Nutrition*, 2002, 132 (11): 3,526S–3,529.

Norat, T., A. Lukanova, P. Ferrari, and E. Riboli. "Meat Consumption and Colorectal Cancer Risk: Dose-Response Meta-Analysis of Epidemiological Studies." *International Journal of Cancer*, 2002, 98: 241–256.

O'Keefe, S. J. D., "The African Way of Life and Colon Cancer Risk." *The American Journal of Gastroenterology*, 2001, 96 (11): 3,220–3,221.

O'Keefe, S. J. D., et al. "Rarity of Colon Cancer in Africans Is Associated with Low-Animal Product Consumption, Not Fiber." *The American Journal of Gastroenterology*, 1999, 94 (5): 1,373–1,380.

Sandhu, M. S., I. R. White, and K. McPherson. "Systematic Review of the Prospective Cohort Studies on Meat Consumption and Colorectal Cancer Risk: A Meta-Analytical Approach." *Cancer Epidemiological Biomarkers & Prevention*, 2001, 10 (5): 439–446.

## Fermented Foods, the Third *F* in Colon Disease Prevention

Aderiye, B. I. and S. A. Laleye. "Relevance of Fermented Food Products in Southwest Nigeria." *Plant Foods for Human Nutrition* (formerly *Qualitas Plantarum*), 2003, 58 (3): 1–16.

Campbell, J. M., L. L. Bayer, G. C. Fahey, et al. "Selected Fructooligosaccharide (1-Ketose, Nystose, and 1F-Beta-Fructofuranosylnystose) Composition of Foods and Feeds." *Journal of Agriculture and Food Chemistry*, 1997, 45: 3,076–3,082.

Galvez, J., M. Rodriguez-Cabezas, and A. Zarzuelo. "Effects of Dietary Fiber on Inflammatory Bowel Disease." *Molecular Nutrition & Food Research*, 2005, 49 (6): 601–608.

Lei, V. and M. Jakobsen. "Microbiological Characterization and Probiotic Potential of Koko and Koko Sour Water, African Spontaneously Fermented Millet Porridge and Drink." *Journal of Applied Microbiology*, 2004, 96 (2): 384–397.

McMaster, L. D., et al. "Use of Traditional African Fermented Beverages as Delivery

Vehicles for *Bifidobacterium lactis* DSM 10140." *International Journal of Food Microbiology*, 2005, 102 (2): 231–237.

O'Keefe, S. J. D., et al. "Why Do African Americans Get More Colon Cancer than Native Africans?" *Journal of Nutrition*, 2007, 137 (1): 175S–182.

Rafter, J. "Probiotics and Colon Cancer." *Best Practice & Research Clinical Gastroenterology*, 2003, 17 (5): 849–859.

Saikali, J., et al. "Fermented Milks, Probiotic Cultures, and Colon Cancer." *Nutrition and Cancer*, 2004, 49 (1): 14–24.

## Soaking Grains to Improve Digestibility

Hotz, C. and R. S. Gibson. "Traditional Food-Processing and Preparation Practices to Enhance the Bioavailability of Micronutrients in Plant-Based Diets." *Journal of Nutrition*, 2007, 137 (4): 1,097–1,100.

Perlas, L. A. and R. S. Gibson. "Household Dietary Strategies to Enhance the Content and Bioavailability of Iron, Zinc and Calcium of Selected Rice- and Maize-based Philippine Complementary Foods." *Maternal & Child Nutrition*, 2005, 1 (4): 263–273.

## Foraged Greens and Folate, the Fourth *F* in Colon Disease Prevention

Burrin, D. G. and B. Stoll. "Emerging Aspects of Gut Sulfur Amino Acid Metabolism." *Current Opinion in Clinical Nutrition & Metabolic Care*, 2007, 10 (1): 63–68.

Giovannucci, E., et al. "Alcohol, Low-Methionine-Low-Folate Diets, and Risk of Colon Cancer in Men." *Journal of the National Cancer Institute*, 1995, 87 (4): 265–273.

Hannon-Fletcher, M. P., et al. "Determining Bioavailability of Food Folates in a Controlled Intervention Study." *American Journal of Clinical Nutrition*, 2004, 80 (4): 911–918.

Mason, J. "Diet, Folate, and Colon Cancer." *Current Opinion in Gastroenterology*, 2002, 18: 229–234.

Su, L. J. and L. Arab. "Nutritional Status of Folate and Colon Cancer Risk: Evidence from NHANES I Epidemiologic Follow-up Study." *Annals of Epidemiology*, 2001, 11 (1): 65–72.

Terry, P., et al. "Dietary Intake of Folic Acid and Colorectal Cancer Risk in a Cohort of Women." *International Journal of Cancer*, 2002, 97 (6): 864–867.

Zhang, S. M., et al. "Folate, Vitamin B$_6$, Multivitamin Supplements, and Colorectal Cancer Risk in Women." *American Journal of Epidemiology*, 2006, 163 (2): 108–115.

## Fats, the Fifth *F* in Colon Disease Prevention

Awad, A., et al. "Peanuts as a Source of Beta-sitosterol, a Sterol with Anticancer Properties." *Nutrition and Cancer*, 2000, 36 (2): 238–241.

Daniel, D. R., et al. "Nonhydrogenated Cottonseed Oil Can Be Used as a Deep Fat Frying Medium to Reduce Trans-Fatty Acid Content in French Fries." *Journal of the American Dietetic Association*, 2005, 105 (12): 1,927–1,932.

Edem, D. O. "Palm Oil: Biochemical, Physiological, Nutritional, Hematological and Toxicological Aspects: A Review." *Plant Foods for Human Nutrition* (formerly *Qualitas Plantarum*), 2002, 57 (3–4): 319–341.

Levi, F., et al. "Macronutrients and Colorectal Cancer: a Swiss Case-Control Study." *Annals of Oncology*, 2002, 13 (3): 369–373.

Schloss, I., et al. "Dietary Factors Associated with a Low Risk of Colon Cancer in Coloured West Coast Fishermen." *South African Medical Journal*, 1997, 87 (2): 152–158.

## Closing the Loop: African American Food Traditions

Anderson-Loftin, W., et al. "Soul Food Light: Culturally Competent Diabetes Education." *The Diabetes Educator*, 2005, 31 (4): 555–563.

Dirks, R. T. and N. Duran. "African American Dietary Patterns at the Beginning of the Twentieth Century." *Journal of Nutrition*, 2001, 131 (7): 1,881–1,889.

Harris, J. B. *Iron Pots and Wooden Spoons*. New York: Simon and Schuster, 1989.

Karanja, N., et al. "Steps to Soulful Living (Steps): A Weight Loss Program for African-American Women." *Ethnicity & Disease*, 2002, 12 (3): 363–371.

Mathews, H. F. "Rootwork: Description of an Ethnomedical System in the American South." *Southern Medical Journal*, 1987, 80 (7): 885–891.

Smith, S. L., et al. "Aging and Eating in the Rural, Southern United States: Beliefs About Salt and Its Effect on Health." *Social Science & Medicine*, January 2006, 189–198.

# Chapter 9: Okinawa Japan: A Cold Spot for Breast and Prostate Cancers

## Experts Interviewed in This Chapter

Barry Halliwell, PhD, E-mail interview on October 29, 2006.
Bruce Trock, PhD, Telephone interview on October 10, 2006.
Bradley J. Willcox, MD, Telephone interview on November 17, 2006.
D. Craig Willcox, PhD, Interview on January 15, 2007.

## Okinawa: A Breast and Prostate Cancer Cold Spot

Jemal, A., et al. "Cancer Statistics, 2006." CA: A Cancer Journal for Clinicians, 2006, 56 (2): 106–130.

Suzuki, M., B. J. Willcox, and D. C. Willcox. "Implications from and for Food Cultures for Cardiovascular Disease: Longevity." Asia Pacific Journal of Clinical Nutrition, 2001, 10 (2): 165–171.

Willcox, B. J., D. C. Willcox, and M. Suzuki. The Okinawa Program. New York: Clarkson Potter, 2001.

## The Natural History of Breast and Prostate Cancer, From Atypical Cell to Carcinoma

Ashbeck, E. L., et al. "Benign Breast Biopsy Diagnosis and Subsequent Risk of Breast Cancer." Cancer Epidemiology Biomarkers & Prevention, 2007, 16 (3): 467–472.

Degnim, A. C., et al. "Stratification of Breast Cancer Risk in Women with Atypia: A Mayo Cohort Study." Journal of Clinical Oncology, 2007, 25 (19): 2,671–2,677.

Hartmann, L. C., et al. "Benign Breast Disease and the Risk of Breast Cancer." New England Journal of Medicine, 2005, 353 (3): 229–237.

Sciarra, A., et al. "Inflammation and Chronic Prostatic Diseases: Evidence for a Link?" European Urology, 2007 52 (4): 964–72.

## Migration Studies and Breast and Prostate Cancers

Deapen, D., et al. "Rapidly Rising Breast Cancer Incidence Rates Among Asian-American Women." International Journal of Cancer, 2002, 99 (5): 747–750.

Nagata, C., et al. "Associations of Mammographic Density with Dietary Factors in Japanese Women." *Cancer Epidemiology Biomarkers & Prevention*, 2005, 14 (12): 2,877–2,880.

Nelson, N. J. "Migrant Studies Aid the Search for Factors Linked to Breast Cancer Risk." *Journal of the National Cancer Institute*, 2006, 98 (7): 436–438.

## Antioxidants from the Primordial Ooze

Halliwell, B. "Oxidative Stress and Neurodegeneration: Where Are We Now?" *Journal of Neurochemistry*, 2006, 97 (6): 1,634–1,658.

Halliwell, B. "Phagocyte-Derived Reactive Species: Salvation or Suicide?" *Trends in Biochemical Sciences*, 2006, 31 (9): 509–515.

Halliwell, B. "Reactive Species and Antioxidants. Redox Biology Is a Fundamental Theme of Aerobic Life." *Plant Physiology*, 2006, 141 (2): 312–322.

Ninfali, P., et al. "Antioxidant Capacity of Vegetables, Spices and Dressings Relevant to Nutrition." *British Journal of Nutrition*, 2005, 93 (2): 257–266.

## Antioxidants for Your Glands

Bettuzzi, S., et al. "Chemoprevention of Human Prostate Cancer by Oral Administration of Green Tea Catechins in Volunteers with High-Grade Prostate Intraepithelial Neoplasia: A Preliminary Report from a One-Year Proof-of-Principle Study." *Cancer Research*, 2006, 66 (2): 1,234–1,240.

Chan, J. M., P. H. Gann, and E. L. Giovannucci. "Role of Diet in Prostate Cancer Development and Progression." *Journal of Clinical Oncology*, 2005, 23 (32): 8,152–8,160.

Duthie, G. and A. Crozier. "Plant-Derived Phenolic Antioxidants." *Current Opinion in Clinical Nutrition & Metabolic Care*, 2000, 3 (6): 447–451.

Higdon, J. V., et al. "Cruciferous Vegetables and Human Cancer Risk: Epidemiologic Evidence and Mechanistic Basis." *Pharmacological Research*, 2007, 55 (3): 224–236.

Ratnam, D. V., et al. "Role of Antioxidants in Prophylaxis and Therapy: A Pharmaceutical Perspective." *Journal of Controlled Release*, 2006, 113 (3): 189–207.

Rock, C. L., et al. "Plasma Carotenoids and Recurrence-Free Survival in Women With a History of Breast Cancer." *Journal of Clinical Oncology*, 2005, 23 (27): 6,631–6,638.

Tuekpe, M., et al. "Potassium Excretion in Healthy Japanese Women Was Increased by a Dietary Intervention Utilizing Home-Parcel Delivery of Okinawan Vegetables." *Hypertension Research*, 2006, 29 (6): 389–396.

Zhang, M., et al. "Green Tea and the Prevention of Breast Cancer: A Case-Control Study in Southeast China." *Carcinogenesis*, 2006: bgl252.

## Anticancer Foods from the Deep

Evert, J., et al. "Morbidity Profiles of Centenarians: Survivors, Delayers, and Escapers." *The Journals of Gerontology Series A: Biological Sciences and Medical Sciences*, 2003, 58 (3): M232–237.

Khotimchenko, I. S and Y. Khotimchenko. "Lead Absorption and Excretion in Rats Given Insoluble Salts of Pectin and Alginate." *International Journal of Toxicology*, 2006, 25: 195–203.

Smyth, P. "The Thyroid, Iodine and Breast Cancer." *Breast Cancer Research*, 2003, 5 (5): 235–238.

Taira, K., et al. "Sleep Health and Lifestyle of Elderly People in Ogimi, a Village of Longevity." *Psychiatry and Clinical Neurosciences*, 2002, 56 (3): 243–244.

Yuan, Y. V. and N. A. Walsh. "Antioxidant and Antiproliferative Activities of Extracts from a Variety of Edible Seaweeds." *Food and Chemical Toxicology*, 2006, 44 (7): 1,144–1,150.

## Antiproliferative Shrooms

Chung, R. "Functional Properties of Edible Mushrooms." *Nutritional Reviews*, 1996, 54: S91–S93.

Cunningham-Rundles, H. L., B. Cassileth. "Are Botanical Glucans Effective in Enhancing Tumoricidal Cell Activity?" *Journal of Nutrition*, 2005, 135: 2919S.

Kodoma N., K. Komuta, H. Nanba. "Can Maitake MD-Fraction Aid Cancer Patients?" *Alternative Medicine Reviews*, 2002, 7: 451.

Sliva., D. "Ganoderma Lucidum (Reishi) in Cancer Treatment." *Integrative Cancer Therapies*, 2003, 2: 358–364.

## The Soy Controversy

Choi, M. S. and K. C. Rhee. "Production and Processing of Soybeans and Nutrition and Safety of Isoflavone and Other Soy Products for Human Health." *Journal of Medicinal Food*, 2006, 9 (1): 1–10.

Lewis, J. G., et al. "Circulating Levels of Isoflavones and Markers of 5[alpha]-Reductase Activity Are Higher in Japanese Compared with New Zealand Males: What Is

The Role of Circulating Steroids in Prostate Disease?" *Steroids*, 2005, 70 (14): 974–979.

Trock, B. J., L. Hilakivi-Clarke, and R. Clarke. "Meta-Analysis of Soy Intake and Breast Cancer Risk." *Journal of the National Cancer Institute*, 2006, 98 (7): 459–471.

## Early Years Matter

Michels, K. B. and W. C. Willett. "Breast Cancer—Early Life Matters." *New England Journal of Medicine*, 2004, 351 (16): 1,679–1,681.

Kehinde, E. O., et al. "Do Differences in Age-Specific Androgenic Steroid Hormone Levels Account for Differing Prostate Cancer Rates Between Arabs and Caucasians?" *International Journal of Urology*, 2006, 13 (4): 354–361.

Shu, X. O., et al. "Soyfood Intake During Adolescence and Subsequent Risk of Breast Cancer Among Chinese Women." *Cancer Epidemiology Biomarkers & Prevention*, 2001, 10 (5): 483–488.

Wu, A. H., et al. "Adolescent and Adult Soy Intake and Risk of Breast Cancer in Asian-Americans." *Carcinogenesis*, 2002, 23 (9): 1,491–1,496.

## More Mojo: Start Low, Go Slow

Bernstein, A. M., et al. "First Autopsy Study of an Okinawan Centenarian: Absence of Many Age-Related Diseases." *Journal of Gerontology Series A: Biological Sciences & Medical Sciences*, 2004, 59 (11): 1,195–1,199.

Willcox, B. J., D. C. Willcox, and M. Suzuki. *The Okinawa Program*. New York: Clarkson Potter, 2001.

## Omega-3s Yet Again

Cui, Y. and T. E. Rohan. "Vitamin D, Calcium, and Breast Cancer Risk: A Review." *Cancer Epidemiology Biomarkers & Prevention*, 2006, 15 (8): 1,427–1,437.

Hirose, K., et al. "Dietary Factors Protective Against Breast Cancer in Japanese Premenopausal and Postmenopausal Women." *International Journal of Cancer*, 2003, 107 (2): 276–282.

Holmes, M. D., et al. "Association of Dietary Intake of Fat and Fatty Acids with Risk of Breast Cancer." *Journal of the American Medical Association*, 1999, 281 (10): 914–920.

Terry, P., et al. "Fatty Fish Consumption and Risk of Prostate Cancer." *The Lancet*, 2001, 357 (9,270): 1,764–1,766.

Wakai, K., et al. "Dietary Intakes of Fat and Fatty Acids and Risk of Breast Cancer: A Prospective Study in Japan." *Cancer Science*, 2005, 96 (9): 590–599.

## Hara Hachi Bu

Heilbronn, L. K. and E. Ravussin. "Calorie Restriction and Aging: Review of the Literature and Implications for Studies in Humans." *American Journal of Clinical Nutrition*, 2003, 78 (3): 361–369.

Sho, H. "History and Characteristics of Okinawan Longevity Food." *Asia Pacific Journal of Clinical Nutrition*, 2001, 10 (2): 159–164.

Willcox, D. C., et al. "Caloric Restriction and Human Longevity: What Can We Learn from the Okinawans?" *Biogerentology*, 2006. 7: 173–177.

## The Right Rice

Hudson, E. A., et al. "Characterization of Potentially Chemopreventive Phenols in Extracts of Brown Rice That Inhibit the Growth of Human Breast and Colon Cancer Cells." *Cancer Epidemiology Biomarkers & Prevention*, 2000, 9 (11): 1,163–1,170.

Panlasigui, L. N. and L. U. Thompson. "Blood Glucose Lowering Effects of Brown Rice in Normal and Diabetic Subjects." *International Journal of Food Sciences and Nutrition*, 2006, 57 (3): 151–158.

## The Okinawan Paradox

Onishi, N. "On U.S. Fast Food, More Okinawans Grow Super-Sized." *New York Times*, March 30, 2004.

# Chapter 10: Foraging for Indigenous Foods in a Modern World

## Experts Interviewed in This Chapter

Dana Friedrich, In-person interview on February 17, 2007.
Brian Halweil, Telephone interview on February 21, 2007.

Paula Jones, Telephone interview on February 23, 2007.
Bill Thompson, Telephone interview on February 22, 2007.

## On Eating Locally

Halweil, B. *Eat Here*. New York: W.W. Norton & Company, 2004.

Michael Pollan, *The Omnivore's Dilemma: A Natural History of Four Meals*. New York: Penguin Press, 2006.

Joel Salatin, *Everything I Want To Do Is Illegal: War Stories From the Local Food Front*. White River Jct, VT: Chelsea Green Publishing, 2007.

Trichopoulou, A., et al. "Traditional Foods: Why and How to Sustain Them." *Trends in Food Science and Technology*, 2006, 17: 498–504.

## Appendices

Berdanier, C. D., J. Dwyer, and E. B. Feldman. *Handbook of Nutrition and Food, Second Edition*. Boca Raton, FL: CRC Press, 2007.

Nestle, M. *What to Eat*. New York: North Point Press, 2006.

Pennington, J. *Bowes and Church's Food Values of Portions Commonly Used*. 18th edition, Philadelphia: Lippincott-Raven, 2005.

Simopoulos, A. and J. Robinson. *The Omega Diet*. New York: HarperCollins Publishers, 1999.

# Acknowledgments

I acknowledge with deep gratitude the dozens of friends, colleagues, and perfect strangers who helped me explore the benefits of indigenous diets. Without their contributions, this project could never have happened.

First and foremost, I want to thank my beloved co-adventurer (and husband) Ross Levy. His patience and support allowed this book to move beyond a mere daydream. I would also like to thank my children, Arlen and Emet Levy, who bravely tasted all the recipes and who never failed to give their honest opinions. I heartily thank Allison Fragakis for her excellent nutrition advice throughout this project and for spearheading the recipe-testing portion of this book.

I am deeply grateful to my parents, Susan and David Miller, whose own wanderlust and love of eating and cooking first launched me into the world of travel adventure and indigenous foods and to my brother Sam Miller, the armchair nutritionist, who seems to know more than many professionals. My in-laws Joy and Jim Caswell also played a vital role with their unfailing encouragement and their generous offers to baby-sit (which allowed me to make my excursions to far-off lands).

A particular thanks must also go to my agent, Diane Bartoli; to my editors at HarperCollins, Jeremy Cesarec, and Mary Ellen O'Neill, as well as my production editor, Amy Vreeland; and to those friends and family who read and critiqued this entire manuscript: Derek Rosenfield, Brigitte Sandquist, Allison Fragakis, Susan Miller, David Miller, Ross Levy, and Elayne Klein. I also thank Ken Howard Wilan, whose expertise and contacts helped me launch this whole project.

I extend my deepest gratitude to my treasured colleagues Avril Swan, Meghan Schwartzman, Gabrielle Modolo, and Jennifer Fullerton, who supported me when I took time off from my clinical work and never voiced anything but enthusiasm for this book project.

Andrew Weil was one of the first people to read parts of this manuscript. His encouragement gave me the will to push on at a time when I was losing steam. For this, and for his wonderful foreword, I am truly grateful.

The dozens of people who served as recipe testers and tweakers must also be acknowledged. Many of them were willing guinea pigs at my dinner table, and others were bold enough to experiment in their own kitchens. The über testers included Yoni Cassel, Karen Drozda, Brigitte Sandquist, Kristi Patterson, Ken Peters, and Dashka Slater. Gail Forrest, Barbara Gross, Christina Nooney, Karen Kasdin, Diana Goldstein, Emily Newman, Gretchen Cammiso, Rebecca Handler, Nina Glatt, Cindy Roberts, Micheal Kabir, Staci Cole, Jane Ryan, Rebecca Katz, Lindsey Cimino, Grant Colfax, Rodman Rogers, and Anne Marie Sarubin also regularly gave their feedback. Thanks to their input, I can say with confidence that the recipes in this book have been tested and met the approval of a wide range of palates.

I am also deeply indebted to the many patients whose stories served as the inspiration for this book. They have been my best teachers, and clearly without them, I would never have conceived of this project.

In addition to the experts mentioned by name throughout the book, I would like to extend a special thanks to individuals in connection with each chapter. They generously gave me hours or days of their time to discuss recipes or cook me a meal, review specific content, arrange interviews, or help me plan travel itineraries. My debt to them is enormous. They include the following, listed by chapter:

## Introduction: The Jungle Effect
Ayla Tiago, Linnea Smith, Peter Jensen, Juvencio Nuñez Pano, and Edemita Peterman Pano

## Chapter 1: Dining in the Cold Spots
Jessica Prentice, Thomas Cowan

## Chapter 2: Anatomy of an Indigenous Diet
Paul Rozin

## Chapter 3: A Diet Lost in Translation
Paul Rozin, Harriet Kuhnlein

## Chapter 4: Feeding Our Genes or Our Taste Buds
Kauila Clark, Jose Ordovas, Robert Nussbaum, and Steve Chen

## Chapter 5: Copper Canyon, Mexico
Amparo Medoza, Jaime Patiño, Jesus Bueno, Laurie Mackenzie, Sharon Meers, Steve Dostart, Judy and Chuck Nichols, and Sandra Slater

## Chapter 6: Crete, Greece
Vassiliki Yiakoumaki, Michael Herzfeld, John and Athene Fragakis, Nikki Rose, Gerald Gass, Penny Manalis, and Antonia Trichopoulou

## Chapter 7: Iceland
Ingur Axelsson, Jóhann Axelsson, Emily Newman, and Jeffrey Jones

## Chapter 8: Cameroon, West Africa
Cori Duncan, David T. Reese, Michel Adjibodou, Genevieve Menanie, Colette Ashby, Dorcas Tchuente, Stephen O'Keefe, Diana Wylie, James McCann, Joseph Holloway, and David Lawrence

## Chapter 9: Okinawa, Japan

Kenji and Chitose Hamada, Masato Goya, Masao Maeshiro, Yaeko Yara, Yuko Onishi, Sumako Yamashiro, Chizuko Yakabe, Ken Shimoji, Amanda Mayer Stinchecum, and D. Craig Willcox

## Chapter 10: Foraging for Indigenous Foods in a Modern World

Paula Jones, Brian Halweil, Dana Friedrich, and Larry Bain

Allison would like to separately acknowledge her superstar husband, Christopher; her cheerleader children, Cassidy and Solomon; her parents, Elayne and Murray Klein; and her in-laws, John and Athene Fragakis.